UNDERSTANDING ULTRASOUND PHYSICS

Second Edition

UNDERSTANDING ULTRASOUND PHYSICS

Fundamentals
and
Exam Review

Second Edition

Sidney K. Edelman, Ph.D.

Program Director
ESP, Inc.

Clinical Instructor
Section of Cardiology
Department of Medicine
Baylor College of Medicine

Woodlands, Texas

ESP, Inc.
PO Box 7439
Woodlands, Texas 77387
(281) 292-9400
FAX (281) 292-9430
www.esp-inc.com

Second Edition, Fourth Printing, 1997.
Second Edition, Third Printing, 1996.
Second Edition, Second Printing, 1994.
Second Edition, 1994.
First Edition, 1990.

Library of Congress Catalog Card Number: 94-70716

ISBN 0-9626444-3-9

Printed in the United States of America by:
Tops Printing, Inc.
Bryan, Texas

With love, to my family:

My wife,
Diane Lynn
and my children,
Lauren Alexis
Jennifer Ruth
and
Reed David

PREFACE

The first edition of this text was published nearly four years ago. Since that time, the fundamentals of ultrasound physics have remained constant. However, the practice of diagnostic ultrasound and this text have undergone dramatic change. The overall structure of the text is similar to that of the first edition. We have stressed brevity for the sake of clarity. However, we have attempted to provide detail regarding important physical principles. Many examples, questions, and quizzes have been added. We believe that this approach enhances the educational process.

This text is designed to help teach the fundamental principles of ultrasound. The book has two major divisions. The first is a discussion of the fundamentals of physics and instrumentation, while the second is a selection of multiple choice questions with detailed answers. This material must be reviewed carefully. Although the text is often brief, each statement is important and should be thoroughly understood. The review questions throughout the text serve to reinforce the subject matter. It is our hope that presenting this material in a variety of formats will clarify some difficult to understand (and difficult to teach) concepts.

In addition, this text was designed to assist sonographers in preparation for certification. The fundamentals and exam review format should help guide most sonographers along the path of professional registration. You will note that there are very few numerical examples in the text. Solving numerical problems is not a required skill for most ultrasound board exams. Rather, devote your energies to understanding concepts and relationships. Try to apply the same skills to the study of ultrasound physics that you use each day as a clinical sonographer. I hope that your chosen profession will be more enjoyable and fulfilling with this knowledge.

SKE, Houston, Texas

ACKNOWLEDGMENT

Most of the credit for this textbook goes to the thousands of sonographers that I have had the pleasure to study with over the past years. They continue to reinforce to me the importance of loyalty, caring, dedication, and perseverance. The great American composer, Arthur Rubenstein, said "I have found that if you love life, life will love you back." It is easy to love ultrasound, because of the professional community -- their generous appreciation, support, and praise. These personal experiences have made my life's work a true labor of love.

I wish to thank my colleagues past, -- the physicians and staff of the Texas Heart Institute and the Clayton Foundation for Research -- who, early in my professional life, instilled in me the need to carefully balance technology and compassionate medical care. And to my colleagues present, many thanks for providing an exciting environment where striving for excellence is an essential part of each and every day.

On a far more personal note, I wish to thank two individuals who have contributed enormously to this effort. I extend my gratitude to Carol Latta for the excellent illustrations that clarify the principles of physics. Sincere appreciation is extended to Rebecca Teaff for her careful and outstanding editorial assistance. As a result of their superb skills, it is easier to *Understand Ultrasound Physics*.

Finally, sincere thanks to my colleague and friend, Leonard Pechacek, RDCS, who introduced me to the remarkable world of diagnostic ultrasound.

CONTENTS

I. FUNDAMENTALS

DEFINITIONS

The term **ULTRASOUND SYSTEM** is used extensively in this text. It refers to the entire system, including the electronics and the transducer. An ultrasound system is changed or altered when a sonographer selects a different transducer.

The following terms are fundamental to the understanding of ultrasound physics:

Unrelated to - two things are not associated or affiliated.

   Ex: Hair color is *unrelated* to shoe size.
          Weight is *unrelated* to eye color.

Related to or **proportional to** - two things are associated or affiliated. However, the relationship is not specified.

   Ex: Taxes are *proportional to* income.
          Santa Claus is *related to* Christmas.
          Exam score is *related to* or *proportional to* studying.

Directly related to or **directly proportional to** - two things are associated, and when one of them increases, so does the other.

   Ex: Clothing size is *directly proportional* to a person's weight.
          Age is *directly related* to experience.
          Skill is *directly related* to practice.
          Quality of wine is *directly related* or *directly proportional* to age.

Inversely related to or **inversely proportional to** - two items are associated but as one of them increases, the other decreases.

   Ex: Golf score is *inversely related* to skill.
          A car's mileage is *inversely proportional* to its weight.
          Grades in school are *inversely related* to or *inversely proportional* to partying time.

| QUESTIONS - RELATIONSHIPS |

Are the following pairs *inversely related*, *directly related*, or *unrelated*?

1. Cholesterol level and longevity.

2. Smoking and the likelihood of cardiovascular disease.

3. Years employed and days of vacation earned per year.

4. IQ and weight.

5. Caloric intake and weight.

6. Hours spent exercising and weight.

7. Alcohol intake and sobriety.

Answers:

1 - Inversely related. As a general rule, the higher an individual's cholesterol level, the shorter is his lifespan.

2 - Directly related. Generally, individuals who smoke are more likely to have cardiovascular disease.

3 - Directly related. The longer a person works for a company, the more vacation he is awarded.

4 - Unrelated. There is no relationship between weight and IQ.

5 - Directly related. The more calories an individual consumes, the greater is his weight.

6 - Inversely related. As people exercise more, their weight decreases.

7 - Inversely related. The more a person drinks, the less sober he becomes.

UNITS

All numerical values should have corresponding units.

For example, the question "How old is Judy?" requires a numerical response with units. If the answer is "6," then "6" alone is an incomplete response. Is it 6 *days, months, years,* or *decades*?

A numerical answer to a question thus requires a unit to be complete.

Units of **length** (distance, circumference) = cm, feet

Units of **area** = cm^2, ft^2

Units of **volume** = cm^3, ft^3, cc's

Units of **time** = seconds, min, years

> To measure the volume of an irregularly shaped object, lower it into a full tank of water. Then remove the object and measure the amount of water that is missing. The volume of the missing water equals the volume of the object.

UNIT CONVERSION

We must know how to change one form of unit to another.

These are forms of unit conversion:
- How many quarters are in 1 dollar?
- How many days are in 1 month?

Note: When we change units, the 'total picture' does not change; only the manner of presentation changes.

- Isn't 12 inches the same as 1 foot?
- Isn't 12 inches the same as 1/3 yard?

RULE: Treat conversion units as fractions with a value of 1 and carry along the units.

Example: Convert 12 inches into centimeters. (Hint: 2.54 cm =1 inch.)

$$12 \text{ inches } \times \left[\frac{2.54 \text{ cm}}{1 \text{ inch}} \right] = 30.48 \text{ cm}$$

↰(This term equals one.)

POWERS OF TEN

Very large or very small numbers are difficult to write.

SCIENTIFIC NOTATION is a shorthand method designed to represent these types of numbers.

RULES: 1) Shift the decimal point so that the resulting number is between 1 and 10. Then,

2) Multiply by the appropriate power of 10.

Examples:

1,000,000	=	1.0×10^6 ☞ this is an exponent
500	=	5.0×10^2
0.79	=	7.9×10^{-1}
0.000124	=	1.24×10^{-4}
1742	=	1.742×10^3

Note: In scientific notation:

- ♦ A number with a POSITIVE exponent is GREATER than 10.

- ♦ A number with a NEGATIVE exponent is LESS than 1.

- ♦ A number with an exponent of zero is between 1 and 10.

METRIC SYSTEM

Powers of Ten	Prefix	Symbol	Meaning
10^9	giga	G	billion
10^6	mega	M	million
10^3	kilo	k	thousand
10^2	hecto	h	hundred
10^1	deca	da	ten
10^{-1}	deci	d	tenth
10^{-2}	centi	c	hundredth
10^{-3}	milli	m	thousandth
10^{-6}	micro	μ	millionth
10^{-9}	nano	n	billionth

COMPLEMENTARY METRIC UNITS

Think of certain pairs as if they "belong together." For example, if we express frequency in units of *MEGA*hertz, then we will refer to a period in units of *MICRO*seconds: millions of cycles and millionths of seconds.

Numbers	Metric Equivalents	
billions and billionths	giga & nano	G & n
millions and millionths	mega & micro	M & μ
thousands and thousandths	kilo & milli	k & m
hundreds and hundredths	hecto & centi	h & c
tens and tenths	deca & deci	da & d

Metric units are used extensively in this text and in all of ultrasound physics. You must learn them!

REVIEW - BASICS

1. Which of the following is an appropriate unit for area?

    a) seconds   b) yards   c) inches

    d) cm^3 e) square yards

2. Which of the following is not an appropriate unit for volume?

    a) cubic miles   b) gallons   c) cm

    d) cm^3 e) pint

3. All of the following are measures of length except:

    a) mile   b) inch   c) mm

    d) second   e) km

4. How many milliliters are in 8 liters?

    a) 1/8   b) 8   c) 80

    d) 800   e) 8000

5. How many centimeters are in 3 meters?

    a) 1/300   b) 1/3   c) 3

    d) 300   e) 3000

6. How many kilometers are in 3000 meters?

    a) 1/300   b) 1/3   c) 3

    d) 300   e) 3000

7. List the following units in increasing order:

    a) mega   b) micro   c) milli

    d) hecto   e) deca   f) deci

8. List the following units in decreasing order:

    a) nano   b) centi   c) giga

    d) kilo   e) hecto   f) micro

9. How are the size of a tree and its age related?

    a) inversely   b) directly   c) unrelated

10. How are the height of a person and the color of his automobile related?

    a) inversely   b) directly   c) unrelated

11. How is the age of a loaf of bread related to its freshness?

    a) inversely   b) directly   c) unrelated

ANSWERS - BASICS

1 - e) square yards: Area is measured in units of length squared. The area that carpet covers is measured in square yards.

2 - c) cm: Volume is measured in units of length cubed. Centimeters is simply a measure of length.

3 - d) second: Seconds are units of time, not length.

4 - e) 8000: Milli means thousandth. There are 1,000 milliliters in 1 liter. Thus, there are 8,000 ml in 8 liters.

> With questions such as this, as the metric prefix gets larger, the number must decrease. When the metric prefix gets smaller, the number must increase.

5 - d) 300: Centi means hundredth. There are 100 centimeters in 1 meter. Thus, there are 300 cm in 3 meters.

6 - c) 3: Kilo means thousand. Three kilometers equal 3,000 meters.

7 - The correct sequence of increasing order is as follows::
      b, c, f, e, d, a   or
      micro, milli, deci, deca, hecto, mega

8 - The correct sequence of decreasing order is as follows:
      c, d, e, b, f, a
      giga, kilo, hecto, centi, micro, nano

9 - b) directly: Generally, as a tree ages, its size increases.

10 - c) unrelated: Generally, a person's age and car color have no relation to with each other.

11 - a) inversely: As a loaf of bread ages, its freshness decreases.

SOUND

ACOUSTIC PROPAGATION PROPERTIES: the effects of the medium upon the sound wave.

BIOLOGIC EFFECTS or **BIOEFFECTS:** the effects of the sound wave upon the biologic tissue through which it passes.

Sound is identified as the rhythmical cycling of acoustic variables.

There are four acoustic variables:

> **Pressure** - concentration of force
> *units*: lb/sq in, Pascals (Pa)

> | Acoustic variables cycle rhythmically throughout time. |

> **Density** - concentration of mass or weight
> *units*: kg/cubic cm

> **Temperature** - concentration of heat energy
> *units*: degrees

> **Distance** - measure of particle motion
> *units*: cm, feet, miles

Transverse wave - particles of the medium move in a direction perpendicular (at right angles) to that of the wave.

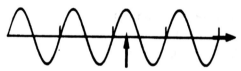

Longitudinal wave - particles of the medium move in the same direction as the wave travels.

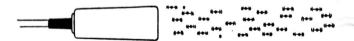

A sound wave is composed of **compressions and rarefactions** in the molecules of the medium. Molecules physically vibrate, indicating that sound waves are mechanical in nature. In order for sound to exist, it must travel through a medium. Sound cannot travel through a vacuum because a vacuum contains no molecules.

Sound is a *mechanical*, *longitudinal* wave that travels in a straight line.

All sound waves are described by the seven parameters listed below:

PERIOD   **PROPAGATION SPEED**
FREQUENCY

AMPLITUDE
POWER
INTENSITY   **WAVELENGTH**

When you complete studying this material, you should understand
why the parameters are arranged in this manner.

PERIOD

PERIOD (of time) is the length of **time** that it takes to complete a single cycle.
For example, the period of the moon circling the earth is 28 days,
and the period of the earth circling the sun is 1 year.

Period may also be described as the time from the start of one cycle to the start
of the next cycle.

Units: seconds, msec, hours -- any unit of time.

Typical values of the ultrasound (US) period in clinical diagnostic imaging are:
0.1 to 0.5 μsec
0.0000001 to 0.0000005 sec
1.0×10^{-7} to 5.0×10^{-7} sec

Period is determined only by the **sound source** and is not affected by the
medium through which the sound travels.

Period *cannot be changed* by the sonographer, as long as the sonographer uses
the same ultrasound system (and transducer).

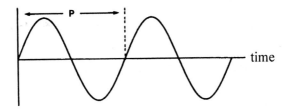

FREQUENCY

FREQUENCY is the number of certain events that occur in a specific
duration of time.

Ex: What is the frequency of presidential elections?
25 times per century
2 1/2 times per decade
1 time every 4 years

In diagnostic ultrasound (US), a wave's frequency is the number of cycles of
an acoustic variable that occur in 1 second.

Units: cycles per second, cycles/second, per second, Hertz, or Hz

HERTZ is another way of saying "per second."

1 cycle/second = 1 Hertz

1,000 cycles/second = 1 kHz

1,000,000 cycles/sec = 1 MHz

Frequency is determined only by the **sound source** (the ultrasound system) and
not by the medium through which the sound travels.

Frequency *cannot be changed* by the sonographer, as long as the sonographer
uses the same ultrasound system (and transducer).

ULTRASOUND is a wave with a frequency exceeding the upper limit of
human hearing, which is 20,000 Hz or 20 kHz. Man cannot hear
ultrasound.

AUDIBLE SOUND is a wave with a frequency between 20 Hz and 20,000
Hz. Sound within this frequency range *can* be heard by man.

INFRASOUND is a wave with a frequency less than 20 Hz. This range of
frequencies is so low that infrasound cannot be heard by man.

Designation	Frequency
Ultra	> 20,000 Hz $(20 \, kHz)$
Audible	between 20 Hz and 20,000 Hz
Infra	< 20 Hz

In clinical imaging, frequencies typically range from 1 MHz to 10 MHz. Frequency affects penetration and axial resolution (image quality).

Eight cycles occur in 4 seconds; the frequency of this wave is 8/4 or 2 Hz.

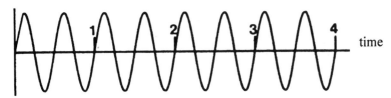

Note: Period and frequency are reciprocals. When any pair of reciprocal numbers are multiplied together, the result is the value 1.

period (sec) x frequency (Hz) = 1

period (sec) = 1 / frequency (Hz)

frequency (Hz) = 1 / period (sec)

Remember to use complementary units for period and frequency; such as sec & Hz, or msec & kHz. (See page 5 for a discussion of complementary metric units).

Frequency and period are inversely related:
    As frequency increases, period decreases.
    As frequency decreases, period increases.

Note: In reality, frequency is determined by the period of a single cycle in the wave. Once the period is known, its reciprocal is the frequency.

Note: Frequency can change only if period changes. Period changes only when frequency changes. If one of these parameters remains constant, then the other remains unchanged.

AMPLITUDE

Three parameters relate to the size of a sound wave. They are **amplitude, power**, and **intensity**. Let us study them separately.

AMPLITUDE relates to the strength of the sound wave. Amplitude describes the magnitude of a wave.

Units: any units of an acoustic variable -
   temperature - **degrees**
   particle motion - **cm, ft**, units of distance
   density - **grams/cubic cm**
   pressure - **pounds/square in, Pascals (Pa)**
   Note: Amplitude may also be reported in special units called **decibels (dB).** A complete description of decibels is found on page 48.

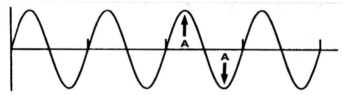

Amplitude equals the difference between the average value and the maximum value of an acoustic variable.

Amplitude is NOT the difference between the minimum and maximum values of an acoustic variable.

Amplitude is initially determined only by the **sound source** (the US system), and not by the medium through which the sound wave travels.

Amplitude decreases as sound propagates through the body.

Amplitude *can be changed* by the operator.

With continuous wave sound (displayed above), the positive and negative amplitudes are equal. With sound pulses, the positive and negative amplitudes are often different (side).

positive amplitude ↘

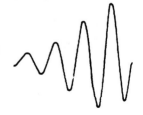

PHASES

IN-PHASE refers to two waves that are in step; in other words, where the maximum and minimum amplitudes of both waves occur at the same instant in time.

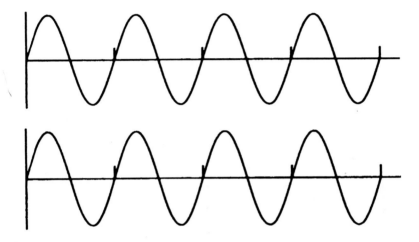

OUT-OF-PHASE refers to two waves that are out of step; in other words, where the maximum (peak) and minimum (trough) amplitudes of both waves occur at different times.

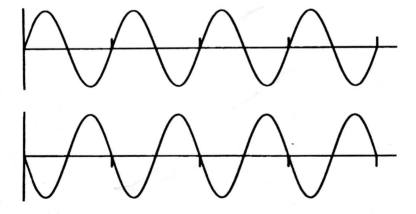

INTERFERENCE OF WAVES

When two waves overlap at the same location and at the same instant in time, they combine with each other. The result of the overlap is the creation of a single, new wave that is the sum of the two original waves. The waves adding together is called **INTERFERENCE**.

CONSTRUCTIVE INTERFERENCE occurs when the amplitude of the new wave is greater than the original two waves. In-phase waves interfere constructively.

DESTRUCTIVE INTERFERENCE occurs when the amplitude of the new wave is less than the original two waves. Out-of-phase waves interfere destructively.

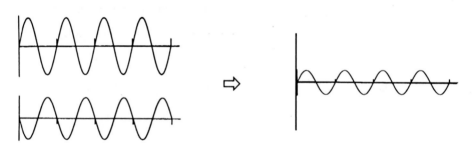

A pair of out-of-phase waves of equal magnitude may destructively interfere with each other and completely cancel themselves out. In effect, both waves vanish.

POWER

POWER relates to the strength of the sound
    wave. It is the rate in which work is
    performed or energy is transferred from
    the entire sound beam.

> Units of POWER: Think of a
> light bulb or a hair dryer!

Units: watts (w)

Power is initially determined only by the **sound source** (the ultrasound system)
    , and not by the medium through which the sound wave travels.

In the clinical setting, power *can be changed* by the operator. When the
    sonographer changes the amplitude of a wave, the power is also
    changed.

Power decreases as ultrasound (US) travels through the body.

Power is directly related to amplitude:
    When amplitude increase, power increases.
    When amplitude decreases, power decreases.

Power is proportional to the wave's AMPLITUDE SQUARED.

$$\text{power} \propto (\text{amplitude})^2$$

    ↰ this term means "proportional to"

Ex: When a wave's amplitude is tripled, the power is increased to 9 times
    its original value. (This is because $3^2 = 9$.)

    If the amplitude is halved, the power is quartered. (Because 1/2
    squared equals 1/4.)

INTENSITY

INTENSITY relates to the strength of the sound beam. It is the concentration of energy in a sound beam.

Intensity equals the power of a beam divided by its cross-sectional area.

Units: watts/cm^2; watts (from power) and cm^2 (from beam area).

The intensity of a beam as it exits a transducer is determined only by the **sound source** (the ultrasound system), not the medium through which the sound travels.

Intensity decreases as ultrasound propagates through the body.

In clinical imaging, since the amplitude and power can be changed by the operator, the intensity can also be changed by the operator.

In clinical imaging, US intensity typically ranges from 0.01 mW/cm^2 to 100 W/cm^2.

$$\text{Intensity (watts / cm sq)} = \frac{\text{power (watts)}}{\text{beam area (cm}^2)}$$

Intensity is directly related to both power and amplitude. When the power or amplitude of a beam increases, so does its intensity. When a beam's power or amplitude decreases, so does the intensity.

Intensity is proportional to the POWER.
If the power is doubled, the intensity is doubled.
If the power is quartered, the intensity is quartered.

Intensity is proportional to the AMPLITUDE of the wave SQUARED.
If the amplitude is doubled, the intensity is increased 4 times.
If the amplitude is quartered, the intensity is reduced to 1/16th of its original value.

Recall that amplitude, power and intensity relate to the strength of an US
beam. When one of these parameters increases, so do the others. A
simple way to remember the
relationships between them is that the
word *squared* always follows the word
amplitude:

> Remember the term
> AMPLITUDE SQUARED.

Power is proportional to the amplitude *squared.*
Intensity is proportional to the amplitude *squared.*
Power is proportional to intensity.

WAVELENGTH

WAVELENGTH is the distance or length that one complete cycle occupies:
the cycle length.

Ex: Similar to the length of a single boxcar in a train of infinite length.

Units: mm, meter -- any unit of length.

Determined by **both the source and the medium.**

Wavelength *cannot be changed* by the sonographer (when the same US system
and transducer are used).

In clinical imaging, US wavelength in soft tissue ranges from 0.1 - 0.8 mm.

$$\text{wavelength(mm)} \; = \; \frac{\text{propagation speed (mm / us)}}{\text{frequency (MHz)}}$$

The detail that is displayed in an image is substantially influenced by
wavelength because wavelength determines longitudinal
resolution. (Longitudinal resolution is discussed on page 90.)

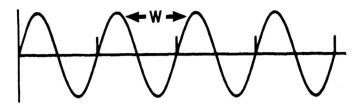

Wavelength is inversely related to frequency:

- Higher frequency waves have shorter wavelengths.
- Lower frequency waves have longer wavelengths.

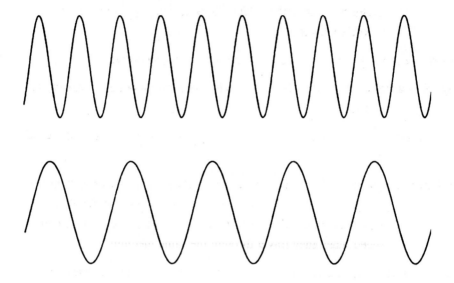

In any medium, as frequency increases, the wavelength decreases.

Shorter wavelengths generally produce higher quality images. This explains why higher frequency transducers create better quality clinical images.

PROPAGATION SPEED

PROPAGATION SPEED is the speed at which sound moves through a
medium -- also called sound's speed, velocity of sound, or
acoustic velocity.

Units: meters per second, mm/μs -- any unit of distance/time.

Speed is determined only by the characteristics of the **medium** -- density and
stiffness.

Propagation speed is not affected by the sound source and *cannot be changed*
by the sonographer.

All sound, regardless of the frequency, travels at the same speed through any
specific medium. This means that sound with a frequency of 5
MHz and sound with a frequency of 3 MHz travel at the same
speed when they travel through the same medium.

In clinical imaging, the speed of sound cannot change unless the medium
through which the wave is traveling also changes.

RULE OF THUMBS:

Density 👍 and Speed 👎 - Opposite letters (D and S) = opposite
directions. As density increases, speed decreases. As density
decreases, speed increases. Therefore, density and speed are
inversely related.

What is density? Density is the concentration of mass per unit volume, or
the weight of 1 cubic centimeter of a material. Steel is dense;
cotton candy is not dense.

Stiffness 👍 and Speed 👍 - Same letters (S and S) = same direction.
As stiffness increases, speed increases. As stiffness decreases, so
does speed. Therefore, stiffness and speed are directly related.

What is stiffness? Stiffness is a material's ability to maintain its shape,
even when a pressure is applied to it. Bone is stiff; lung tissue is
not.

Note: Acoustic velocity is determined by a **combination of density and stiffness.**

Materials that are very stiff but not dense will have the fastest speed. (Bone is very stiff and not particularly dense. Therefore, it has a high speed.)

Materials that are very dense but not stiff will have the slowest speed. (Air has an exceedingly low stiffness and, therefore, a low speed.)

Note: Compressibility and elasticity are the opposite of stiffness.

Rule of thumbs works for these terms also. Materials with greater compressibility or elasticity have lower speeds.

> Generally, velocities are lowest in gases, higher in liquids, and highest in solids.

The average speed of sound in biologic "soft tissue" is as follows:

1.54 km/s, 1,540 m/s, 1.54 mm/μs

Propagation speeds: air < lung <   fat < soft tissue < bone

  (mm/μs)   .33   0.5   1.45   1.54   4.0

Propagation speed (m/s) = frequency (Hz) **x** wavelength (m)

TISSUE	SPEED
Air	330 m/s
Muscle	1,580 m/s
Liver	1,550 m/s
Kidney	1,560 m/s
Brain	1,520 m/s
Soft tissue	1,540 m/s

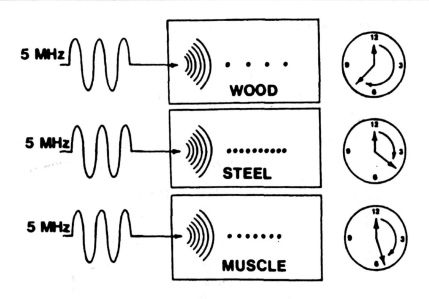

Acoustic waves travel at unequal speeds though different media. The
propagation speed depends upon the density and elastic properties
of each medium.

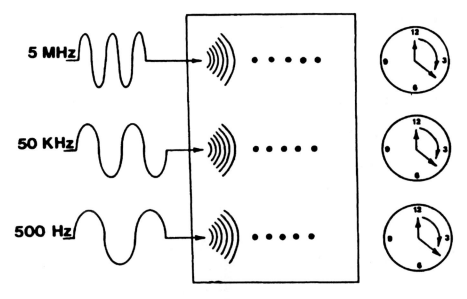

All acoustic waves, even those with different frequencies, travel at the same
speed through a particular medium. In a single medium, all sound
waves travel at the same speed.

PARAMETERS OF CONTINUOUS WAVES

PARAMETERS	BASIC UNITS	UNITS	DETERMINED BY
period	time	sec, usec	sound source
frequency	1/time	1/sec, Hz	sound source
amplitude	acoustic variables	lb/in², dB, etc	sound source
power	work/time	watts	sound source
intensity	power/area	watts/sq cm	sound source
wavelength	distance	mm, cm	source & medium
propagation speed	distance/time	m/sec	medium

Amplitude and, therefore, power and intensity can be changed by the operator. The values of these parameters decrease as sound propagates through the body.

REVIEW - SOUND

1. Give the units of measure for each term below.
   a) wavelength   b) frequency   c) intensity
   d) propagation speed   e) period   f) power

2. Which determines each of these parameters, the medium or the US system?
   a) wavelength   b) frequency   c) intensity
   d) propagation speed   e) period   f) power

3. Which of the following cannot be changed by the sonographer?
   a) wavelength   b) amplitude   c) intensity
   d) propagation speed   e) period   f) power

4. True or False? Sound is a transverse, mechanical wave.

5. True or False? A wave with a frequency of 15,000 MHz is ultrasonic.

6. True or False? If the amplitude of a wave is increased to 3 times its
   original value, the intensity is increased by 6 times.

7. True or False? If the power of a wave is halved, the intensity is reduced to
   one-fourth its original value.

8. True or False? Propagation speed increases as frequency increases.

9. Which of the following are considered acoustic variables?
   a) frequency   b) density   c) particle motion
   d) temperature   e) period   f) pressure

10. How many cycles occur in 0.5 seconds if the frequency is 5 million Hz?

11. Two sound beams with different frequencies are traveling through the
    same medium. Are the following parameters larger, smaller, or
    the same for the wave with the higher frequency?
    a) wavelength   b) propagation speed
    c) period   d) amplitude

12. The effects of ultrasound upon tissue are called _____ .

13. What is the acoustic velocity in soft tissue?

14. Medium 1 has a density of 9 and a stiffness of 6.
    Medium 2 has a density of 8 and a stiffness of 6.
    Which medium has a lower propagation speed?

ANSWERS - SOUND

1 -   a) millimeters   b) Hertz   c) watts/cm^2
      d) meters/sec   e) second   f) watts

2 -   a) both.   b) sound source   c) sound source
      d) medium   e) sound source   f) sound source

3 -   a) cannot   b) can   c) can
      d) cannot   e) cannot   f) can

4 - False: Sound is mechanical, but it is a longitudinal wave.

5 - True: Ultrasound is defined as any wave with a frequency exceeding
           20,000 hertz. In this example, the wave's frequency is
           15,000 MHz or 15,000 million hertz.

6 - False: Intensity is proportional to the amplitude squared. If amplitude is
           tripled, the intensity is increased by a factor of nine.

7 - False: Intensity is proportional to the power. If the power is divided in
           half, the intensity will also be half as great.

8 - False: Propagation speed is determined only by the medium. Frequency
           does not affect the propagation speed.

9 - b, c, d, and f: Density, particle motion, temperature, and pressure.

10 - 2,500,000 cycles

11 -   a) smaller   b) same
       c) smaller   d) same

12 - biologic effects

13 - In soft tissue, sound travels 1,540 m/s, or 1.54 km/sec. It is very
     important to know this value.

14 - Medium #1 has the lowest propagation speed. Since the stiffnesses of the
     two media are the same, the medium with the GREATER density
     has the LOWER propagation speed.

REVIEW - SOUND AND RELATIONSHIPS

1. All of the following may be units of amplitude except:
             a) millimeters   b) degrees   c) dB
             d) Pascals   e) m/s

2. How are the following parameters related - directly, inversely, or unrelated?
   a)   Frequency and period.
   b)   Amplitude and power.
   c)   Amplitude and intensity.
   d)   Power and intensity.
   e)   Wavelength and intensity.
   f)   Wavelength and frequency.
   g)   Acoustic velocity and density.
   h)   Elasticity and speed of sound.
   i)   Acoustic velocity and compressibility.
   j)   Stiffness and sound speed.
   k)   Frequency and sound speed.
   l)   Frequency and intensity.
   m)   Power and frequency.

ANSWERS - SOUND AND RELATIONSHIPS

1 - e) m/s: Meters per second are units of speed. Speed is not an acoustic
            variable. Therefore, its units cannot describe an amplitude.

2 -   **Relationship**
   a)   inversely.
   b)   directly.
   c)   directly.
   d)   directly.
   e)   unrelated.
   f)   inversely.
   g)   inversely.
   h)   inversely.
   I)   inversely.
   j)   directly.
   k)   unrelated.
   l)   unrelated.
   m)   unrelated.

PULSED ULTRASOUND

In diagnostic ultrasound, continuous waves cannot create an image. Rather, short bursts. or pulses, of acoustic energy are produced and then received by the transducer to construct an image.

A **PULSE** of ultrasound is a collection of cycles that travel together. Consider a pulse to be analogous to a train. Although there are individual cycles (cars) that make up the pulse (train), the pulse moves as one. Wherever the locomotive leads, the other cars follow.

Note: A pulse *must have* a beginning and an end; otherwise it is a continuous wave.

Components of a pulse:

> the cycles ("transmitting," "on," or "talking" time)
> dead time ("receiving," "off," or "listening" time)

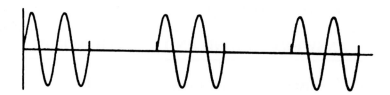

All sound waves are described by the seven parameters (described on pages 9 through 22. Five additional parameters describe pulsed sound waves.

Five additional descriptors of pulsed ultrasound:

Pulse duration	Duty factor
Pulse repetition period	Spatial pulse length
Pulse repetition frequency	

PULSE DURATION

PULSE DURATION is the *time* from the START of a pulse to the END of
that pulse -- ONLY the time that the pulse is "on" or transmitting.

Units: seconds, msec -- any unit of time.

Pulse duration is determined by the **source** only.

It is determined by the *number of cycles* in the pulse (the amount of 'ringing')
and the *period* of each cycle.

Short pulses have the following traits:
- ◆ Few cycles (less ringing).
- ◆ Each individual cycle has a short period (higher frequency).

The number of cycles in a pulse is determined by the manufacturer's
design of the transducer.

The period of each cycle is determined by the frequency of the sound.

Pulse duration is a characteristic of an US system and remains the same as long
as the system components remain unchanged. A pulse is a pulse is a
pulse! Pulse duration *cannot be changed* by the operator.

In clinical imaging, pulse
duration ranges from
0.5 to 3 μsec.

Generally, shorter pulses create
higher quality images.

> Pulse duration will increase if:
> 1) Period of the cycles increases,
>   requiring a lower frequency.
> 2) # cycles in the pulse increase,
>   requiring a new transducer design.

Pulse duration (msec) = # cycles in pulse **x** period (msec)

Pulse duration (msec) = # cycles in pulse / frequency (kHz)

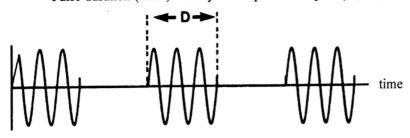

time

PULSE REPETITION PERIOD

PULSE REPETITION PERIOD (PRP) is the *time* from the START of one
pulse to the START of the next pulse. This includes both the time
that the pulse is "on" (the pulse duration) and the "dead time."

The PRP is analogous to "period."

Units: seconds, minutes -- any unit of time.

PRP is made up of the both the pulse duration and the time that the system is
listening.

PRP is determined only by the **sound source** (the ultrasound system) and not
the medium through which the sound travels.

Pulse repetition period *can be changed*
by the sonographer. When the
sonographer adjusts the
maximum imaging depth (depth
of view), he alters the receiving
time. Remember that the
receiving time is part of the
pulse repetition period.

As imaging depth increases,
PRP increases.
When depth decreases,
PRP decreases.

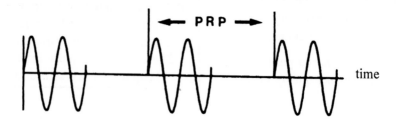

The depth of view and pulse repetition period are directly related:

> When a sonographer increases depth of view, the listening time increases, and therefore PRP increases.

> When the sonographer decreases the depth of view, the listening time decreases, and therefore the PRP decreases.

> Note: The sonographer may alter the PRP and does so by changing only the listening time. The sonographer cannot alter the pulse duration.

In clinical imaging, the PRP has values from 100 μsec to 1 msec.

PULSE REPETITION FREQUENCY

PULSE REPETITION FREQUENCY (PRF) is the number of pulses that occur in a single second, or the number of times the transducer is excited each second. (We don't care about the characteristics of the pulse. We are only measuring the number of pulses that are created in each second!)

PRF is analogous to frequency.

PRF is determined only by the **source**.

> As imaging depth increases, PRF decreases.

Units: Hz, per second, units of frequency.

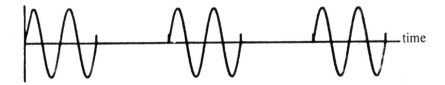

PRF *can be changed* by the sonographer. When the sonographer alters the maximum imaging depth (depth of view), the PRF changes.

When an ultrasound system images deeper, the receiving or "listening" time increases, thereby reducing the PRF.

In clinical imaging, the PRF ranges from 1,000 to 10,000 Hz (1-10 kHz).

PRF is inversely related to depth of view:
- As maximum imaging depth increases, the PRF decreases.
- As maximum imaging depth decreases, the PRF increases.

Pulse repetition period and pulse repetition frequency are reciprocals. They are inversely related. Both the PRP and PRF depend upon imaging depth.

pulse repetition period (sec) **x** pulse repetition frequency (Hz) = 1

$$\text{pulse repetition period (sec)} = \frac{1}{\text{pulse repetition frequency (Hz)}}$$

$$\text{pulse repetition frequency (Hz)} = \frac{1}{\text{pulse repetition period (sec)}}$$

Use complementary metric units for the PRF and the PRP.
Ex: PRF in kHz & PRP in msec or PRF in Hz & PRP in sec.

In most cases, only one pulse of US is allowed to travel in the body at a time. Therefore, the PRF changes as the imaging depth changes. The operator determines the maximum imaging depth, and thus alters the pulse repetition frequency.

DUTY FACTOR

The **DUTY FACTOR** is the *percentage* or *fraction of time* that the US machine is producing a pulse or transmitting sound. Duty factor is also called *duty cycle*.

Maximum value = 1.0 or 100%

UNITLESS!

Minimum value = 0.0 or 0%

$$\text{Duty Factor } (\%) = \frac{\text{pulse duration (sec)}}{\text{pulse repetition period (sec)}} \times 100$$

Duty factor is determined only by the **source,** not by the medium.

Duty factor *can be changed* by the sonographer.

Duty factor is important in relation to intensity. See page 39 for a complete discussion of intensities.

In clinical imaging, the duty factor ranges from 0.001 to 0.01. Ultrasound systems spend only a small percentage of time transmitting sound and a majority of time listening for reflected pulses.

As we know, the operator adjusts the maximum imaging depth by adjusting the listening time and, thereby, determines the pulse repetition period. Thus, while adjusting the imaging depth, the operator also changes the duty factor.

While all other parameters remain constant, if the:
1) PRF increases, then the duty factor increases.
2) maximum imaging depth increases, then the duty factor decreases.
3) pulse repetition period increases, then the duty factor decreases.
4) pulse duration increases, then the duty factor increases. (Note: The only way pulse duration can be changed is by using a different US system. For example, by selecting a different transducer).

> Any action that increases the percentage of time an ultrasound system is transmitting sound will increase the duty cycle.

If the duty factor is 1.0 or 100%, then the machine is always producing a pulse. It is a continuous wave system.

If the duty factor is 0.0, then the machine is never producing a pulse. It is off.

SPATIAL PULSE LENGTH

SPATIAL PULSE LENGTH (SPL) is the length or distance that a pulse
occupies in space. SPL is the distance from the start of a pulse to
the end of that pulse.

Units: mm, meter -- any unit of distance.

Spatial pulse length is analogous to wavelength.

Ex: The overall length of our imaginary train from the front of the locomotive
to the end of the caboose.

The SPL will decrease when:
- the wavelength decreases (higher frequency sound).
- there are fewer cycle in the pulse (new transducer design).

SPL is determined by **both** the **source** and the **medium**.

In clinical imaging, spatial pulse length ranges from 0.1 to 1 mm.

SPL *cannot be changed* or adjusted by the sonographer. Spatial pulse length is
changed only when a different system or transducer is used.

Spatial pulse length is important in that it determines longitudinal resolution
(image quality).

spatial pulse length (mm) = # of cycles in pulse **x** wavelength (mm)

If we know the length of each boxcar in a train and the number of cars in the
train, then we know the total length of the train.

Spatial pulse length is:
- inversely proportional to the frequency.
- directly proportional to wavelength.
- directly proportional to # cycles in the pulse.

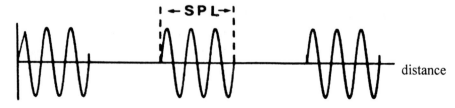

PARAMETERS THAT ARE THE SAME FOR PULSED AND CONTINUOUS WAVES

PERIOD	**FREQUENCY**	These parameters describe features of a single cycle and are not affected by whether the wave is continuous or pulsed.
WAVELENGTH	**PROPAGATION SPEED**	

PARAMETERS OF PULSED WAVES

Parameters	Basic Units	Units	Determined By	Common Values
pulse duration	time	sec, msec	sound source	0.5-3.0 μsec
pulse repetition period	time	sec, μsec	sound source	0.1-1.0 msec
pulse repetition frequency	1/time	1/sec, Hz	sound source	1-10 kHz
spatial pulse length	distance	mm, cm	source & medium	0.1-1.0 mm
duty factor	none	none	sound source	0.001-0.01

When adjusting the maximum imaging depth, the operator changes the listening time, which affects the pulse repetition period, pulse repetition frequency, and duty factor.

The pulse duration and the spatial pulse length are constant for a particular ultrasound system and cannot be changed by the sonographer. They are independent of listening time.

PULSED ULTRASOUND - GENERAL CONCEPTS

1. Pulsed ultrasound is required to make images. Continuous wave ultrasound cannot make images.

2. Short sound pulses create higher quality images with greater detail. Therefore, US manufacturers strive to create systems that produce pulses with short pulse durations and short spatial pulse lengths. This is accomplished by damping the transducer's ceramic and decreasing the amount of 'ringing'.

3. The wavelength of a high-frequency cycle is short. Therefore, pulses made of high-frequency sound tend to be short. This is why higher frequency transducers create high-quality images.

4. US manufacturers design and fabricate transducers to produce pulses made of very few cycles. This tends to creates short pulses and thereby optimize image quality.

5. The pulse duration and spatial pulse length of a particular transducer can never be changed by the sonographer. (Hint: A pulse is a pulse is a pulse!)

6. When the sonographer alters the maximum imaging depth (also called depth of view), the only characteristic that actually changes is the system's listening time for returning echoes.

7. When the sonographer adjusts maximum imaging depth, the only parameters that are altered are those that incorporate the receiving time, the pulse repetition period, pulse repetition frequency, and duty factor.

REVIEW - PULSED WAVES

1. _____ is the time from the start of a pulse to the end of that pulse.

2. _____ is the time from the start of a pulse to the start of the next pulse.

3. Pulse repetition frequency is the reciprocal of _____.

4. What are the duty factors of the following?

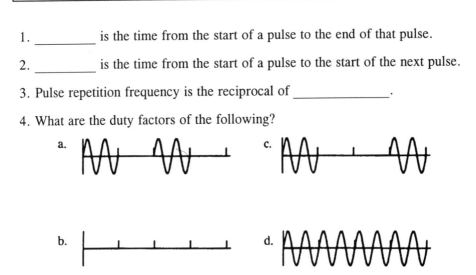

5. By changing the imaging depth, which of the following parameters does the operator also change?

    a) pulse repetition frequency   b) duty factor   c) propagation speed

    d) pulse repetition period   e) amplitude   f) spatial pulse length

6. The speed of a 5 MHz continuous wave is 1.8 km/sec in a particular medium. If the wave is then pulsed with a duty factor of 0.5, calculate the new propagation speed?

7. Which pair of patterns could be produced by the same transducer?

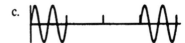

ANSWERS - PULSED WAVES

1 - Pulse duration

2 - Pulse repetition period

3 - Pulse repetition period

4 - a. 0.5   b. 0.0   c. 0.333   d. 1.0

> Use the following steps to determine a duty factor:
>   a) Determine the "on" time, called the pulse duration.
>   b) Determine the "off" time.
>   c) Add the "on" and "off" times together. This is the PRP.
>   d) Divide the pulse duration by the pulse repetition period.
>   This is the duty factor.

5 - By changing the imaging depth, the operator also changes the following
       parameters:
       a) pulse repetition frequency
       b) duty factor
       d) pulse repetition period

6 - The propagation speed depends only upon the medium through which the
       sound travels. The new propagation speed is exactly the same as
       the old propagation speed, 1.8 km/sec.

7 - A single transducer may create pulses B and C. The pulses themselves
       are identical. Only the time between pulses is different. Recall
       that a particular transducer always creates identical pulses.

REVIEW - PULSED WAVES

1. Which one of the following parameters can be changed by the sonographer?
   a) pulse duration   b) pulse repetition period c) wavelength
   d) frequency   e) spatial pulse length

2. All of the following are determined solely by the US system and not the medium except:
   a) pulse duration   b) frequency   c) spatial pulse length
   d) period   e) power

3. How are the following variables related (directly, inversely, unrelated)?
   a)   Listening time and depth of view.

   b)   Spatial pulse length and the number of cycles in the pulse.

   c)   Spatial pulse length and frequency of the cycles in the pulse.

   d)   Spatial pulse length and pulse duration.

   e)   Maximum imaging depth and spatial pulse length.

   f)   Maximum imaging depth and pulse repetition period.

   g)   Maximum imaging depth and pulse repetition frequency.

   h)   Pulse repetition period and pulse repetition frequency.

   i)   Image quality and spatial pulse length.

   j)   Image quality and pulse duration.

   k)   Wavelength and spatial pulse length.

ANSWERS - PULSED WAVES

1 - b) pulse repetition period: PRP comprises the transmit and receive
        times. The sonographer adjusts the receive time and therefore
        adjusts the PRP.

2 - c) spatial pulse length: The SPL is determined by both the source and
        the medium

3 - a) Directly: The longer an US system listens, the deeper is its maximum
        imaging depth.

   - b) Directly: Cycles are the building blocks of a pulse. The more
        cycles, the longer the pulse.

   - c) Inversely: Higher frequency cycles have short wavelengths. Thus,
        the higher the frequency, the shorter the spatial pulse length.

   - d) Directly: Pulses that are long are long both in distance and in
        duration.

   - e) Unrelated: Maximum imaging depth is determined by listening time.
        Spatial pulse length is related only to the transmission of a pulse or
        the "on" time. Thus, depth and SPL are unrelated.

   - f) Directly: The deeper an US system images, the longer the listening
        time. PR period contains both talking and listening time. Thus,
        when depth increases, so does PRP.

   - g) Inversely: The deeper a system images, the fewer pulses are emitted
        per second.

   - h) Inversely: PRP and PRF are reciprocals. As one increases, the other
        decreases.

   - i) Inversely: Higher quality images are created by shorter pulses.

   - j) Inversely: Higher quality images are created by shorter pulses.

   - k) Directly: Cycles are the building blocks of pulses. If the wavelength
        of the cycles are long, so too is the pulse.

INTENSITY

INTENSITY is the concentration of the power in a beam.

Intensity = power / beam area

Units: watts/square cm, W/cm^2

Intensity is important in the study of bioeffects because the ultrasonic dose must be known in order to analyze harmful effects.

All ultrasound beams experience variations in intensity based on the location within the beam where the measurement is made. Beams are not uniform in space.

Ultrasound pulses have large variations in intensity, depending on whether the system is transmitting or receiving. Pulsed ultrasound beams are not uniform throughout time.

Therefore, it is possible to characterize US intensities in several ways with regard to space and time.

Four key words:

Spatial - referring to distance or space.
Temporal - referring to time.
Peak - the maximum value.
Average - the mean value.

Four ways to measure intensity:

SPTP intensity - spatial peak, temporal peak.
SATP intensity - spatial average, temporal peak.
SPTA intensity - spatial peak, temporal average.
SATA intensity - spatial average, temporal average.

Spatial - an ultrasound beam does not have the same intensity at different
    locations within the beam.

Temporal - a pulsed ultrasound beam does not have the same intensity at
    different times.

In addition to SPTP, SATP, SPTA, and SATA intensities, there are two
    additional intensities, only for pulsed ultrasound, that warrant
    discussion:

SPPA intensity - spatial peak, pulse average.

SAPA intensity - spatial average, pulse average.

Pulse average (PA) intensity is the average intensity that occurs in a beam
    only within the time that the pulse exists. The terms including
    pulse average have become part of the description of intensities
    because the intensity of a short ultrasound pulse varies throughout
    the pulse duration.

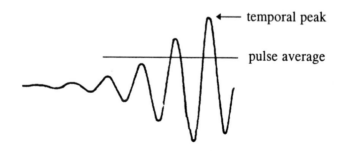

As we know, the peak intensity that appears within a pulse is named the **temporal peak intensity**. The intensity averaged over the duration of the pulse only (the "on" time) is called the **pulse average** intensity. The intensity averaged over the entire pulse repetition period (the "on" time and the "off" time") is the **temporal average intensity**.

Temporal Peak (TP) is the maximum at any instant in time.

Pulsed Average (PA) is the average throughout only the pulse duration (only the transmitting time).

Temporal Average (TA) is the average throughout all time (both transmitting and receiving).

Units: W/cm^2. (Every intensity will have these units!)

SAPA has a value that is between SATP and SATA.

SPPA has a value that is between SPTP and SPTA.

I_m is the average intensity of a sound beam calculated only during the most intense half cycle. Typically, I_m is similar in value to the SPTP intensity.

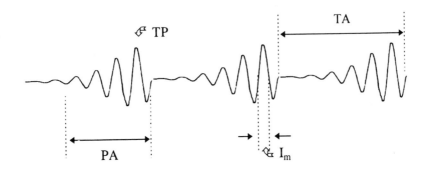

BEAM UNIFORMITY COEFFICIENT

BEAM UNIFORMITY COEFFICIENT (BUC) - also called the **SP/SA factor** - is a unitless ratio that describes the distribution of an ultrasound beam in space. It is the spatial peak intensity divided by the spatial average intensity.

$$\frac{SP}{SA} \; Factor \;\; = \;\; \frac{spatial \; peak \; intensity \; (W \,/\, cm^2)}{spatial \; average \; intensity \; (W \,/\, cm^2)}$$

> BUC is associated with space.

The SP/SA factor is analogous to the duty factor, except it relates to space rather than time. Since the SP intensity must exceed the SA intensity, the SP/SA factor has:

- a minimum value of 1.0
- a maximum value greater than 1.0 **UNITLESS!**

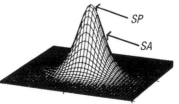

Note: The closer the SP/SA factor is to 1, the more even or homogeneous the beam is. The larger the SP/SA factor, the brighter is the center of the beam compared with its edges.

DUTY FACTOR

The duty factor is a unitless number with a value between 0 and 1; it is the percentage of time the sound pulse is "on."

> Duty factor is associated with time.

For a discussion of the duty factor, see page 30.

$$temporal \; average \; intensity \;\; = \;\; pulsed \; average \; intensity \; \textbf{x} \; duty \; factor$$

> In physics, it is common that parameters with the term *factor* or *coefficient* are unitless. This isn't always true, but it is useful as a general rule.

CONVERTING FROM ONE INTENSITY TO ANOTHER

Rules to remember regarding intensity conversion:

1. Separate the process into two parts: the spatial and then the temporal.

2. The duty factor has values between 0 and 1.

3. The SP/SA factor has values greater than or equal to 1.

4. Peak intensity is always greater than or equal to average intensity (SPTP is always largest; SATA is smallest).

5. To convert the spatial component (e.g., SP to SA), use the SP/SA factor.

6. To convert the temporal component (e.g., PA to TA), use the duty factor.

7. When any number is multiplied by another number that is greater than 1, the resulting number is larger than the original number.

   Ex. $5 \times 4 = 20$ $0.15 \times 4 = .6$

8. When any number is multiplied by another number that is less than 1, the resulting number is smaller than the original number.

   Ex. $10 \times 0.5 = 5$ $0.5 \times 0.2 = 0.1$

9. When any number is divided by another number that is greater than 1, the resulting number is smaller than the original number.

   Ex. $10 / 2 = 5$ $9 / 3 = 3$

10. When any number is divided by another number that is less than 1, the resulting number is greater than the original number.

    Ex. $5 / .5 = 10$ $9 / 0.1 = 90$

EXAMPLES - INTENSITIES

The SPPA intensity of a pulsed wave is 4 mW/cm^2, the duty factor is 0.002, and the SP/SA factor is 6. Find the SPTA and the SATA intensities.

A. Converting the SPPA intensity to the SPTA intensity.

In order to convert the temporal component (PA to TA) while the spatial component (SP) remains the same, use the duty factor (Rule #6). For pulsed waves, TA is always less than PA and therefore, the end result will be a lower intensity. Thus, the SPPA intensity must be multiplied by the duty factor (Rule #8).

SPPA x duty factor = SPTA

4   x   0.002   =   0.008 mW/cm^2

B. Converting the SPTA intensity to the SATA intensity.

In order to convert the spatial component (SP to SA) while the temporal component (TA) remains the same, use the beam uniformity coefficient (Rule #5). SA is always less than SP and, therefore, the end result will be a lower intensity. Thus, the SPTA intensity is divided by the beam uniformity factor (Rule #9).

SPTA / (SP/SA)   = SATA

0.008 / 6   = 0.00133 mW/cm^2

SUMMARY - INTENSITIES

The subject of intensities is quite complex. The following summary may be used to guide the student and sonographer to the correct answer to many questions on this subject.

1. Intensities are reported in various ways. This is important to remember in regard to the bioeffects of ultrasound.

2. The most relevant intensity in the study of bioeffects is the SPTA intensity because it correlates with the heating of tissues. (See page 226 for a discussion of bioeffects.)

3. All intensities have units of watts/cm^2.

4. The SPTP intensity has the highest value. The SATA has the lowest.

5. Generally, the intensities rank in the following order, from largest to smallest:

    SPTP I_m SPPA SPTA SATP SAPA SATA

6. For continuous wave US, the temporal peak and temporal average intensities are the same because the beam is always "on." Thus,
    SPTP = SPTA   and   SATP = SATA

7. The beam uniformity coefficient (BUC), or SP/SA factor, describes the distribution of a wave's intensity over space. Specifically, it is the distribution of intensity through the beam's cross-sectional area. The BUC is unitless and has a value of 1 or more.

8. The duty factor describes the relationship between pulse average ("on" time only) and temporal average ("on" and "off" times) intensities. The duty factor is unitless and has a value from 0 to 1.

9. TP intensity is maximum in time.
    I_m intensity is average intensity during the largest 1/2 cycle.
    PA intensity is average during the pulse duration ("on" time).
    TA intensity is average during the PRP ("on" and "off" times).

10. SP intensity is maximum in space.
    SA intensity is averaged over the cross-sectional area of the beam.

REVIEW - INTENSITIES

The SATP intensity is 1 mW/cm², the duty factor is 1.2 msec, and the SP/SA
factor is 5 percent.

1. Which parameters have incorrect units? mW/cm²

2. Which parameters have correct units? 5%

3. Which parameters have numerical values that are impossible? 0-1

4. Which intensity will be the lowest? SATP

5. Which intensity will be the highest? SPSP

6. Is this pulsed or continuous wave US?

7. If this were a continuous wave, which intensities would be the same?

ANSWERS - INTENSITIES

1 - The SATP intensity should have units of mW/cm^2.

   The duty factor is unitless.

2 - The SP/SA factor has correct units.

> The word **percent** is not a "unit." It simply means hundredths. For example, if a man states that his weight has dropped 10%, we do not know how much weight he have lost. All that is known is that one-tenth of his total body weight has been shed. Therefore, a number followed by a "percent" is merely a ratio; it is a unitless number.

3 - The duty factor must be between 0 and 1.

   The SP/SA factor must be greater than 1.

4 - SATA is always the lowest intensity.

5 - SPTP is always the highest intensity.

6 - It cannot be determined whether this is pulsed or continuous wave ultrasound. The duty factor must be known to make this determination, and that information is erroneous. If the duty factor is 1, then continuous wave sound exists. When the duty factor is between 0 and 1, pulsed sound exists.

7 - SATP and SATA are the same for continuous wave ultrasound. SPTP and SPTA are the same for continuous wave ultrasound.

   Temporal peak and temporal average intensities are the same for continuous wave sound waves because the beam is always present, and there is never a listening time.

LOGARITHMS

A **LOGARITHM** represents an alternative or new way to rank numbers.

> The Richter scale used to measure earthquakes is logarithmic.

The log of any number represents the number of times "10" must be multiplied together to create the original number.

Ex: What is the log of 100? We have to multiply 10 by itself *2 times* to create the number 100, so the log of 100 is 2.

   10 x 10 = 100   Therefore, log 100 = 2

   What is the log of 1000? We have to multiply 10 by itself *3 times* to create the number 1000, so the log of 1000 is 3.

   10 x 10 x 10 = 1000   Therefore, log 1000 = 3

DECIBELS

A **DECIBEL (dB)** is a unit to express a relative difference between two acoustic signals. Decibels are based on a logarithmic scale.

A relative scale. A decibel is a ratio of the final to the initial intensity levels. Two intensities are required to calculate decibels. For a chart of decibels and intensity ratios, see page 242.

Decibels are useful when comparing numbers. For example, dB can be used to describe the following:
- the amplitude of a sound wave (see page 12).
- the degree of amplification of an electrical signal.
- the degree of weakening of a wave as it travels (see page 51).

> When a "ratio" of two intensities, powers, or amplitudes is reported, the units are likely to be dB.

Positive decibels means that the final intensity exceeds the original intensity. The signal is strengthened.

3 dB means a doubling of intensity. It means that the final intensity is twice as big as the original intensity.

6 dB means a double doubling, or, the final is 4 times the original. (2 x 2 = 4)

9 dB means a double, double, doubling. The final intensity is 8 times the original. (2 x 2 x 2 = 8)

10 dB means that the final intensity is 10 times bigger than the original.

Negative decibels means that the final is less than the original intensity.

-3 dB means the intensity has fallen to 1/2 the original value.

-6 dB means the intensity has fallen to 1/4 (1/2 of 1/2) of the original value.

-9 dB means the intensity has fallen to 1/8 (1/2 of 1/2 of 1/2) of the original.

-10 dB means the intensity has fallen to 1/10 of the original value.

Note: Decibel units are used to relate two values to each other. For example, when the sonographer increases the power produced by an ultrasound transducer by 3 dB, the power has doubled from its original value.

What is the original power?   It is unknown.
What is the new power?   It is unknown.

The relationship between the two power levels is known. Thus, units of decibels are used to compare two values to each other.

REVIEW - DECIBELS

1. True or False? Decibels are useful when comparing numbers. ⊤

2. Positive decibels means that the intensity _____ (is increasing, is decreasing, remains the same).

3. Negative decibels means that the intensity _____ (is increasing, is decreasing, remains the same).

4. A reduction of intensity to 1/2 its original value is ___-3___ dB.

5. A reduction in intensity to 1/4 its original value is ___-6___ dB.

6. -10 dB means that the intensity is reduced to __1/10__ of its original value.

7. dB is a mathematical representation with a _____ scale.
   a) multiplication   b) division
   c) longitudinal   d) logarithmic

8. True or False? We need one intensity to calculate decibels.

9. A wave's initial intensity is 2 mW/sq cm. There is an increase of 9 dB. What is the final intensity?
   a) 6 mW/cm cubed   b) 2 mW/cm sq
   c) 16 mW/cm sq   d) 16 μW/cm sq

10. Every 3 dB means that the intensity will __double__.

11. Every 10 dB means that the intensity will _____ .

12. If the final intensity is more than the initial intensity, then the gain in dB is ___+___ (+ or -).

13. If the initial intensity is less than the final intensity, then the gain in dB is ___+___ (+ or -).

ANSWERS:

1 - True.	2 - is increasing.	3 - is decreasing.
4 - -3 dB.	5 - -6 dB.	6 - 1/10.
7 - logarithmic.	8 - False.	9 - c) 16 mW/cm sq.
10 - double.	11 - increase ten-fold.	12 - positive.
13 - positive.		

ATTENUATION

ATTENUATION is the weakening of a sound wave as it travels in a medium.

> Attenuation is a toll that all waves must pay to propagate.

Attenuation is the decrease in intensity and amplitude as sound travels; it is the compromise that sound makes to travel through the medium.

Units: dB, decibels (must be negative, since the intensity is decreasing).

Note: When reporting attenuation in dB, authors will often omit the negative sign. For example, they will simply state "6 dB of attenuation." Since attenuation is the weakening of a beam, the dB's must be negative. Thus, the meaning is -6 dB.

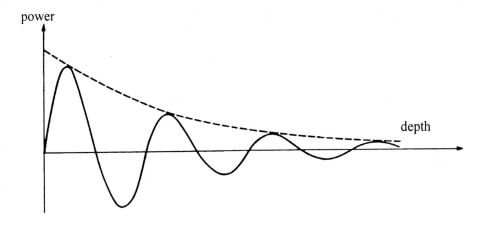

Overall attenuation is determined by both the following:
- Frequency of the sound.
- The distance that the pulse travels.

The further ultrasound travels, the greater the attenuation. Distance and attenuation are directly related.

In soft tissue, the greater the frequency, the greater the attenuation. Frequency and attenuation are directly related.

In the clinical setting, the amount of attenuation determines the frequency of
the transducer that is selected. When a sonographer images to a
substantial depth, a lower frequency transducer must be used. The
sonographer sacrifices image quality for penetration. The absolute
maximum amount of attenuation occurs with the use of high-
frequency sound that travels a long distance.

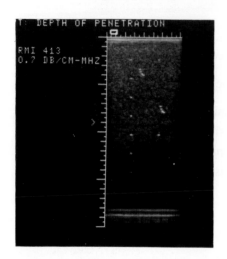 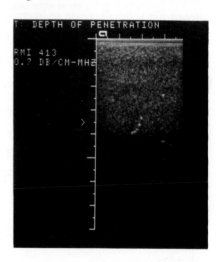

Low frequency transducer penetrates to a greater depth (left) but has a lesser
image quality than a high frequency transducer (right). (Source: Gammex RMI)

Attenuation results from three processes:

ABSORPTION occurs when the ultrasonic energy imparted to biologic tissues
is lost by its conversion to another form of energy, such as heat or
the mechanical vibration of intracellular particles.

Absorption is directly related to US frequency:
* Higher frequencies undergo greater absorption.
* Lower frequencies undergo lesser absorption.

In soft tissue, absorption is the primary component of attenuation.
It is estimated that 80% or more of the total attenuation in soft
tissue is caused by absorption.

REFLECTION Sound often changes direction when it reaches a boundary
between two dissimilar media. Reflection occurs when some of
the propagating acoustic energy is redirected and returns toward

the transducer. Reflections off of a very smooth reflector (such as a mirror) are called **specular**. Reflections act to weaken the sound beam that continues to propagate deeper into the tissues.

SCATTERING If the boundary between two media has irregularities, then the wave is distributed in a number of different directions. Also called **diffuse scattering**, these echoes are much weaker than specular reflections produced at boundaries.

Scattering occurs when a sound wave strikes material whose size is approximately equal to or smaller than the wavelength of the cycles in the pulse. Substantial scattering occurs in lung tissue because the alveoli of the lung are filled with air.

Scattering is directly related to US frequency:
- Scattering increases with higher frequencies.
- Scattering decreases with lower frequencies.

RAYLEIGH SCATTERING is a special form of scattering. When a reflector is much smaller than the sound's wavelength, the ultrasonic energy is diverted in all directions. The higher the frequency, the greater the amount of Rayleigh scattering. In clinical imaging, the interaction of ultrasound and red blood cells results in Rayleigh scattering.

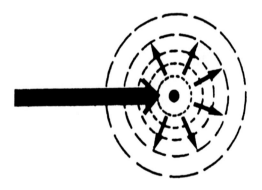

Attenuation ultimately limits the maximum imaging depth achieved during clinical study.

Attenuation in soft tissue is directly related to the frequency of the ultrasound.
- Higher frequencies attenuate more.
- Lower frequencies attenuate less.

Thus, frequency and maximum imaging depth are inversely related:
- Higher frequencies have a shallower imaging depth.
- Lower frequencies have a deeper imaging depth.

ATTENUATION COEFFICIENT

The **ATTENUATION COEFFICIENT** is the amount of attenuation PER CENTIMETER of tissue.

> As frequency increases, the attenuation coefficient increases.

It is the toll that sound pays per centimeter to travel.

Units: dB/cm, decibels per centimeter.

Attenuation coefficient is directly related to frequency:
- Higher frequencies have higher attenuation coefficients.
- Lower frequencies have lower attenuation coefficients.

IN SOFT TISSUE:

The higher the frequency, the greater the attenuation coefficient.

> A reasonable estimate of the attenuation coefficient of sound in soft tissue is half the frequency.

In clinical imaging, it is difficult to determine the attenuation coefficient because this requires measuring intensity within the body. Attenuation coefficient ranges from 0.5 - 1.1 dB/cm per MHz of frequency.

Generally, Atten Coeff (dB/cm) = frequency (MHz)/2

Compared with that in soft tissue, the attenuation coefficient is higher in both bone and lung. US cannot effectively penetrate into these tissues.

Bone is a great absorber.

Lung is a great scatterer.

Note: The attenuation coefficient does not change as sound travels in soft tissue.

total atten (dB) = path length (cm) **x** attenuation coefficient (dB/cm)

HOW TO FIND FINAL INTENSITY OF ULTRASOUND TRAVELING THROUGH SOFT TISSUE

1. Attenuation coefficient (dB/cm) = frequency (MHz)/2.

2. Attenuation coefficient (dB/cm) x path length (cm) = attenuation (dB).

3. Find the intensity ratio in the chart in the reference section on page 242.

4. Multiply original intensity by the intensity ratio.

PROBLEM: ATTENUATION OF TRAVELING SOUND

In soft tissue, an 8 MHz ultrasonic wave starts with an intensity of
$$15 \text{ mW/cm}^2.$$
> 1. What is the intensity at a depth of 2.5 cm?
> 2. What is the intensity at 0.75 cm?

Solution:

1. Path length = 2.5 cm

    Attenuation coefficient = frequency/2 = 8/2 = 4 dB/cm

    Total attenuation = 4 x 2.5 = 10 dB

    > Recall that attenuation is a weakening, or decrease, in the intensity of a beam. Thus, 10 dB of attenuation actually means -10 dB.

    Intensity ratio   = 0.1   (from chart on page 242, one-tenth remains.)

    Final intensity   = 15 mW/sq cm X 0.1

    = 1.5 mW/sq cm, at a depth of 2.5 cm

2. Path length = 0.75 cm

    Attenuation coefficient = frequency/2 = 8/2 = 4 dB/cm

    Total attenuation = 4 x 0.75 = 3 dB

    Intensity ratio = 0.5, there is one-half remaining

    Final intensity   = 15 mW/sq cm X 0.5

    = 7.5 mW/sq cm at a depth of 0.75 cm

HALF VALUE LAYER THICKNESS

The **HALF VALUE LAYER THICKNESS** is the thickness of tissue required
to reduce the intensity of a sound beam by one half. This
represents a -3 dB drop in intensity, or 3 dB of attenuation.

It is also known as: **depth of penetration, half boundary layer**, or
penetration depth.

Units: cm -- any unit of distance.

In clinical imaging, the half value layer thickness ranges from 0.3 to 1 cm.

Penetration depth is inversely related to frequency:

- Higher frequencies produce shallower penetration depths.
- Lower frequencies produce deeper penetration depths.

$$\text{penetration depth (cm)} = \frac{3}{\text{attenuation coefficient (dB / cm)}}$$

> In soft tissue, the half boundary thickness
> is smaller when the:
> 1) attenuation coefficient increases.
> 2) frequency increases.

The half value layer thickness depends upon both the sound's frequency and
the medium through which the sound is propagating.

Ex: Estimate the half value layer thickness of 4 MHz ultrasound in soft tissue.

thickness (cm) = 3/attenuation coefficient (dB/cm)

attenuation coefficient = frequency/2 = 4/2 = 2 dB/cm

thickness = 3/2 = 1.5 cm

Thus, a 4 MHz sound beam will attenuate to half of its intensity for every
1.5 cm that it travels in soft tissue.

IMPEDANCE

IMPEDANCE is the acoustic resistance to sound traveling through a medium.

It is represented by the letter "Z".

Units: RAYL

Impedance is a characteristic of the **medium** only. Thus, each tissue has its own acoustic impedance.

impedance (rayls) = density (kg/m^3) **x** propagation speed (m/s)

Density is a property of the medium through which sound travels.

Propagation speed is a property of the medium.

Therefore, impedance is a property of the medium only.

> A medium has a high impedance when
> * it is very dense and/or
> * it has a fast propagation speed.

Note: Impedance is not directly measured, but rather it is derived from a calculation. To determine a tissue's impedance, the density and propagation speed are measured. Then, these numbers are multiplied together.

Impedance of biologic tissues ranges from 1,250,000 to 1,750,000 Rayls. (1.25 - 1.75 MRayls).

Impedances are important in the physics of *reflection*.

> Reflection of an ultrasound wave depends upon a difference in the acoustic impedances of the two tissues at a boundary.

REVIEW - ATTENUATION, HALF VALUE LAYER THICKNESS AND IMPEDANCE

1. Name the three elements of attenuation.

2. Which of the three elements contributes most to total attenuation in soft tissue?

3. When sound travels, as the distance increases, the attenuation of US in soft tissue _____.

4. Attenuation in lung tissue is _____(less than, greater than, the same as) soft tissue.

5. Attenuation in bone is _____ soft tissue.

6. What are the units of attenuation?

7. What is the relationship between ultrasound frequency and the attenuation coefficient in soft tissue?

8. For ultrasound with an attenuation coefficient of 1 dB/cm in soft tissue, what is the penetration depth?

9. What are the units of the half value layer thickness?

10. As frequency decreases, depth of penetration _____.

11. As path length increases, the half boundary layer _____.

12. Impedance is a characteristic of _____.

13. As the path length increases, the attenuation coefficient of US in soft tissue _____.

14. Acoustic impedance = _____ X _____

15. Impedance is important in _____ at boundaries.

ANSWERS - ATTENUATION, DEPTH OF PENETRATION AND IMPEDANCE

1 - Absorption, reflection and scattering.

2 - Absorption is the primary component of attenuation in soft tissue.

3 - increases.

4 - greater than (because of scattering)

5 - greater than (because of absorption

6 - decibels (dB)

7 - In soft tissue, the attenuation coefficient in dB per centimeter is
         approximately equal to half of a sound wave's frequency in MHz.

8 - penetration depth (cm) = 3 / attenuation coefficient
         penetration depth (cm) = 3/1 or 3 centimeters.
      Solution: The beam will lose half of its intensity every
                  3 cm it travels in soft tissue.

9 - The units are those of distance - centimeters.

10 - increases

11 - remains the same

12 - only the medium

13 - remains the same

14 - Impedance = density (kg/m^3) X propagation speed (m/s)

15 - reflections

REFLECTIONS AND INCIDENCE

SPECULAR REFLECTIONS are produced when an ultrasound wave strikes a smooth surface. In a smooth surface, the irregularities in the surface are large when compared with the sound beam's wavelength. The reflection is well-defined and regular, similar to the reflection produced by a mirror.

SCATTERING (also called **non-specular**) is the random redirection of sound waves in multiple directions. Scattering is produced when ultrasound waves strike a rough surface, where the surface irregularities are the same size as or less than the wavelength. Scattering is also produced when sound waves strike random, small particles or structures located in tissues.

Scattering is directly related to frequency. Higher frequency sound scatters more. Less scatter occurs with lower frequencies.

RAYLEIGH SCATTERING is a form of scatter occurring when sound scatters systematically in all directions. This occurs when sound strikes a red blood cell or small intracellular particles. Rayleigh scattering also occurs when the size of the particle is much smaller than the wavelength.

> Note: All forms of scatter, including back and Rayleigh, produce much weaker echoes when compared with specular reflections.

BACKSCATTER (also called **diffuse scattering**) is scatter returning in the general direction of the transducer.

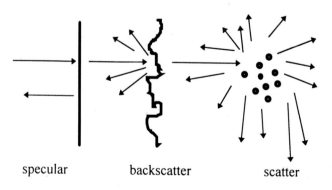

        specular   backscatter   scatter

ANGLE TYPES

There are three types of angles: acute, right and obtuse.

ACUTE ANGLE - any angle of less than 90 degrees.

RIGHT ANGLE - any angle having a measure of exactly 90 degrees.

OBTUSE ANGLE - any angle having a measure greater than 90 degrees.

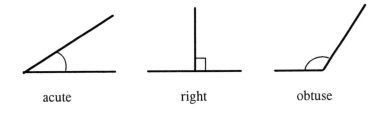

acute   right   obtuse

INCIDENCE

NORMAL INCIDENCE occurs when the sound beam strikes the boundary between different tissues at **exactly 90 degrees** (also called right angles, perpendicular, orthogonal).

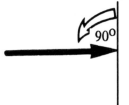

OBLIQUE INCIDENCE occurs when the sound beam strikes a boundary between tissues at an angle other than 90° (not at right angles, not 90 degrees). Both acute and obtuse angles are oblique.

REFLECTION AND TRANSMISSION

INCIDENT INTENSITY is the intensity of the wave that is just about to
strike an interface or boundary between two different media.

REFLECTED INTENSITY is the intensity that, after striking a boundary
between dissimilar tissues, changes direction and returns back in
the direction from which it came.

TRANSMITTED INTENSITY is the intensity that, after striking a boundary,
continues traveling in the same general direction.

Units: W/cm^2 (ALL intensities have these units!)

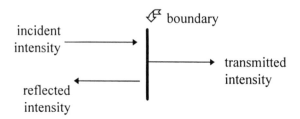

incident (starting) intensity = reflected intensity + transmitted intensity

In clinical imaging, only a very small percentage (1% or less) of the intensity
is reflected back toward the transducer at a boundary between soft
tissues. The remainder is transmitted and continues to propagate in
the forward direction.

A much larger percentage (30%-50%) of a wave's intensity is reflected at a
boundary between bone and soft tissue.

Nearly total reflection occurs at an interface between air and soft tissue.

INTENSITY REFLECTION COEFFICIENT (IRC) is the *percentage* of the
ultrasound (US) intensity that is bounced back when the sound
beam passes from one tissue to another.

INTENSITY TRANSMISSION COEFFICIENT (ITC) is the *percentage* of
the ultrasound (US) intensity that is allowed to pass through when
the beam reaches an interface between two media. The wave
continues to propagate in the forward direction.

Both IRC and ITC: range from 0% to 100%   **Unitless!**

range from 0 to 1.0

100% = intensity reflection coefficient + intensity transmission coefficient

At the boundary between two tissues:

      1) If the IRC and ITC are added together, the result is 100%.

      2) If the reflected and transmitted intensities are added together, the sum equals the incident intensity.

There is "conservation of intensity" at a boundary.

When there is total *reflection* of a wave at a boundary, the intensity reflection coefficient is 100% and the intensity transmission coefficient is 0%.

When there is total *transmission* of a wave at a boundary, the intensity reflection coefficient is 0% and the intensity transmission coefficient is 100%.

REFLECTION WITH NORMAL INCIDENCE

When a wave strikes a boundary between two media **at 90° (normal incidence)**, reflection occurs only if there are *different acoustic impedances* between the two media.

To determine whether reflection occurs when a sound wave strikes an interface at normal incidence, simply ask the question: "Are the acoustic impedances of the tissues on either side of the boundary different?"

> An echo must be produced in order to create an ultrasound image of an anatomic structure. With normal incidence, an impedance difference is required to generate an echo.

In clinical imaging, only 1% or less of the incident energy is reflected back toward the transducer at an interface bounded by soft tissues.

An impedance difference causes reflection of a portion of an incident sound wave at a boundary. The size of the reflection is determined by the *difference in impedances* of the two media comprising the boundary. The more substantial the difference in impedance, the greater the size of the echo.

With normal incidence between the wave and the boundary of the media, the intensity reflection coefficient can be determined using either of the following equations:

$$\text{Intensity reflection coefficient } (\%) = \frac{\text{reflected intensity } (w/cm^2)}{\text{incident intensity } (w/cm^2)} \times 100$$

$$\text{Intensity reflection coefficient } (\%) = \left[\frac{Z_2 - Z_1}{Z_2 + Z_1} \right]^2$$

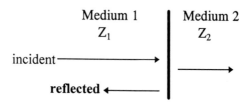

TRANSMISSION WITH NORMAL INCIDENCE

The **INTENSITY TRANSMISSION COEFFICIENT** is the **percentage** of the
ultrasound intensity that is allowed to pass through when the beam
strikes a boundary between two media.

Ranges from 0% to 100%
Unitless!
Ranges from 0 to 1.0

with NORMAL incidence:

INTENSITY TRANSMISSION - when the two media have the identical
impedances, then all of the intensity is transmitted. None is
reflected.

Inten. Transmission Coeff. = 100% - intensity reflection coefficient

$$= \frac{\text{transmitted intensity } (w / cm^2)}{\text{incident intensity } (w / cm^2)} \times 100$$

If the IRC is 90%, then the ITC is 10%. Their sum must equal 100%.

If the IRC is 0%, then the ITC is 100%, and there is total transmission. With
normal incidence, this occurs when the media on both sides of the
boundary have identical impedances.

In clinical ultrasound imaging, 99% or more of the incident energy is
transmitted forward at a boundary between soft tissues.

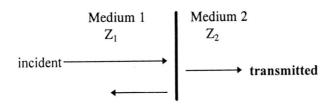

REFLECTION AND TRANSMISSION WITH OBLIQUE INCIDENCE

Extremely complex physics defines transmission and reflection with obliquity. Oblique incidence means that the angle between the boundary and the sound wave is different from 90 degrees.

> With oblique incidence, we cannot predict whether transmission and/or reflection will occur.

Transmission and reflection MAY or MAY NOT occur with oblique incidence, but there are no "simple" rules to describe the process.

The **INCIDENT ANGLE** is the angle between the incident sound beam and the interface between tissues (also called the angle of incidence).

The **REFLECTION ANGLE** is the angle between the reflected sound and the interface between tissues (also called the angle of reflection).

When reflection does occur:

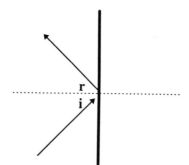

reflection angle = incident angle

This is how a rear-view mirror is used in driving an automobile.
When the interface is smooth, these are specular reflections.
Please note: This relationship is NOT Snell's Law! (See page 67.)

REFRACTION

REFRACTION is a phenomenon associated with transmission.

Refraction is a change in direction, or a bending away from a straight line path, of a wave traveling from one medium

> Refraction causes a straight straw that sits in a glass of water to appear bent or broken.

to another. The wave continues in the same overall direction, but is slightly deflected.

Refraction only occurs when there are
- different **PROPAGATION SPEEDS** and
- **OBLIQUE INCIDENCE** between the sound wave and the boundary.

> The physics of refraction is described by SNELL'S LAW.

Refraction cannot occur with the following:
- Normal incidence or
- Identical sound speeds.

The amount of the deflection is directly related to the change in acoustic velocity from one tissue to the other:
- Vastly different velocities create substantial deviations.
- Slightly different velocities create small deviations from a straight line path.

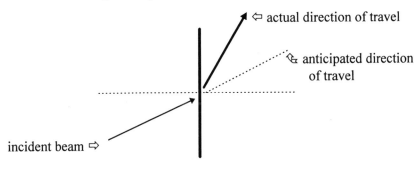

⇐ actual direction of travel

⇐ anticipated direction of travel

incident beam ⇨

To determine whether refraction occurs, simply ask these two questions:
1. "Are the propagation speeds different?" and
2. "Is the oblique incidence?"

SNELL'S LAW defines the physics of refraction:

$$\frac{\text{sine transmission angle}}{\text{sine incident angle}} = \frac{\text{propagation speed of medium 2}}{\text{propagation speed of medium 1}}$$

What is a "sine"?

Every angle has a "sine" associated with it. Refer to the chart of angles and their respective sines on page 242.

When propagation speed 2 is greater than propagation speed 1, then the transmission angle is greater than the incident angle.

When propagation speed 2 is less than propagation speed 1, then the transmission angle is less than the incident angle.

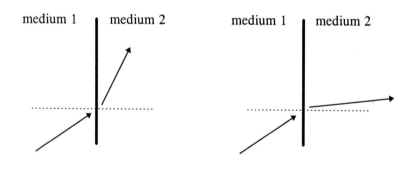

medium 1   medium 2   medium 1   medium 2

prop. speed 1 < prop. speed 2   prop. speed 1 > prop. speed 2

this symbol means "less than"   this symbol means "more than"

EXAMPLE - ATTENUATION, REFLECTION AND REFRACTION

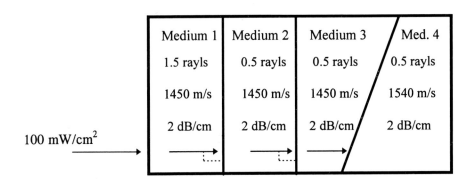

1.   What does the 100 mW per square cm stand for?

2.   What does the 2 dB/cm represent?

3.   If the media are soft tissue, what is a reasonable estimate of the sound's frequency?

4.   What property has units of RAYLS? How is it determined?

5.   What type of incidence is between media 1 and 2?

6.   What occurs at the boundary between media 1 and 2? Why?

7.   What can occur at the boundary between media 2 & 3? Why?

8.   What cannot occur at the boundary between media 2 & 3?

9.   What type of incidence is between media 3 and 4?

10. What can occur at the boundary between media 3 and 4? Why?

11. What process occurs as the US passes through all media? What are the units of this process?

ANSWERS - ATTENUATION AND REFLECTION

1 -   The incident intensity of the ultrasound beam.

2 -   The attenuation coefficient of the media.

3 -   4 MHz. In soft tissue, the attenuation coefficient is approximately half
       of the frequency (In other words, the frequency of sound is
       approximately twice the attenuation coefficient.)

4 -   Impedance. Impedance = density x propagation speed.
       It cannot be measured. It is calculated.

5 -   Normal, perpendicular, 90 degrees, right angles, orthogonal.

6 -   Reflection and transmission. There is normal incidence and
       different acoustic impedances.

7 -   Transmission only. The media's impedances are the same.

8 -   Refraction - because the incidence is normal. Reflection - because the
       impedances are the same with orthogonal incidence.

9 -   There is oblique incidence between media 3 and 4.

10 -   Refraction - because there are different propagation speeds and
       oblique incidence.
       Transmission and reflection may also occur with oblique incidence,
       but this is not certain. The physics is too complex to predict.

11 -   Attenuation (scattering, absorption and reflection). dB.

RANGE EQUATION

How does an imaging system determine the depth of a reflecting surface?

US systems do not measure distance with a ruler. Rather, the "time of flight" of an ultrasound pulse is used to calculate the depth of a reflector.

TIME-OF-FLIGHT, also called **go-return time**, is the elapsed time between pulse production and echo reception by the transducer. In ultrasonic imaging, the go-return time is very short because sound travels nearly 1 mile per second in soft tissue.

Estimating distance from the go-return time is called **echo ranging**. The greater the go-return time, the deeper the reflecting surface.

Since the average speed of US in soft tissue (**1.54 km/sec**) is known, the time of flight and distance that US travels are directly related. The mathematical relationship is called the range equation:

$$\text{depth} = \text{velocity of sound x} \ \frac{\text{time - of - flight}}{2}$$

In soft tissue:

$$\text{depth (mm)} = 0.77 \ \textbf{x} \ \text{go-return time (us)}$$

In soft tissue, sound travels 1 inch in 16 millionths of a second!

In soft tissue, sound travels to and returns from a depth of 1 cm in 13 millionths of a second.

When a pulse's "time of flight" is known, then the distance to a reflector can be calculated. This is how ultrasound systems measure distance.

Time-of Flight	Depth of Reflector	Distance Traveled
13 μs	1 cm	2 cm
26 μs	2 cm	4 cm
39 μs	3 cm	6 cm
52 μs	4 cm	8 cm
130 μs	10 cm	20 cm
260 μs	20 cm	40 cm

A pulse can go to and return from a reflector positioned 20 cm deep nearly four times in one thousandth of a second. (Sound "boogies" in the body!)

TRANSDUCERS

A **TRANSDUCER** is any device that converts one form of energy into another
(e.g., acoustic (sound) to electrical; electrical to thermal).

electric motor - converts electrical energy into mechanical
light bulb   - converts electrical energy into heat and light
muscle   - converts chemical energy into mechanical energy
loudspeaker   - converts electrical energy into acoustic energy

Ultrasound transducers convert electrical energy into acoustic energy during
transmission and acoustic to electrical energy during reception.

PIEZOELECTRIC EFFECT - a property of certain materials to create a
voltage when they are mechanically deformed.
Also, piezoelectric materials change shape and vibrate when a
voltage is applied to them (occasionally, this is called the "reverse
piezoelectric effect.")
This is how ultrasound transducers work. Their primary
component is a crystal with piezoelectric properties.

Piezoelectric Materials: also called *ferroelectric*.

Found in nature: quartz, Rochelle salts, tourmaline
Man-made: barium titanate, lead metaniobate, lead titanate,
lead zirconate titanate (**PZT**)

CURIE TEMPERATURE - If we heat piezoelectric material above this
temperature, the material is *depolarized* and loses its
piezoelectricity forever! We cannot heat sterilize transducers.

SINGLE CRYSTAL TRANSDUCER ARCHITECTURE

Active element - the piezoelectric crystal itself. Lead zirconate titanate (PZT) is the most commonly found piezoelectric material in diagnostic imaging transducers.

Damping element (backing material) - a material that is bonded to the back side of the active element and acts to limit the "ringing" of the crystal. It shortens the pulse duration and the spatial pulse length. The damping material acts to improve the image quality (see p. 90). It also decreases the quality factor and the sensitivity of the transducer. Backing material is commonly made of epoxy resin impregnated with tungsten.

Damping material:
- shortens the spatial pulse length and pulse duration.
- improves picture quality.
- increases bandwidth (the range of frequencies within the pulse).
- decreases the "Q" factor.
- decreases the transducer's sensitivity to reflected echoes

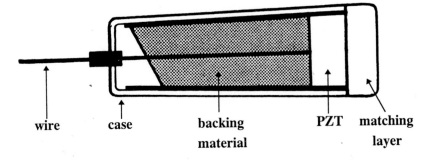

wire   case   backing material   PZT   matching layer

Matching layer - Recall that large differences in impedance result in big echoes at a boundary. PZT has an impedance that is much greater than that of skin. By itself, this would create a large reflection at the skin, and very little acoustic energy would be transmitted into the body. This would limit the performance of ultrasound in imaging tissues within the body.

The matching layer has an impedance between those of the skin and the active element, so more sound energy is transmitted between the crystal and skin.

Acoustic gel has an impedance between those of the matching layer and the skin to further facilitate transmission of acoustic energy. Impedances in decreasing order are:

PZT ⇨ matching layer ⇨ gel ⇨ skin

Wire - Each active element in a transducer requires electrical contact so that the electrical pulse from the ultrasound system can excite the crystal and produce an ultrasonic wave.

Similarly, during reception, the activity or vibration of the crystal produces a voltage. The voltage is transmitted back to the ultrasound system via the wire for further processing into an image.

Case and Insulator - a plastic or metal housing surrounding the above components. This housing:
- prevents electrical noise from interfering with transducer's performance.
- protects the fragile components of transducer.
- protects the patient from electrical shock.

TRANSDUCER FREQUENCIES

How is the frequency of the sound produced by the piezoelectric crystal of a
transducer determined?
The answer is different for continuous wave and pulsed wave
transducers.

For continuous wave ultrasound, the sound's frequency is determined by the
electrical frequency of the excitation voltage applied to the active
element by the US system. The frequency of the continuous wave
sound is identical to that of the continuous electrical signal that
drives the PZT crystal.

5 MHz electrical   5 MHz acoustic

For pulsed ultrasound, the pulse repetition frequency (PRF) is determined by
the ultrasound system (see page 29.) The number of electrical
spikes that the system delivers to the active element equals the PRF.

The FREQUENCY of the sound is determined only by the
characteristics of the PZT crystal.

The frequency of sound emitted by a pulsed wave transducer is
determined by a combination of two factors:
   ♦ the thickness of the crystal.
   ♦ the propagation speed of sound in the crystal.

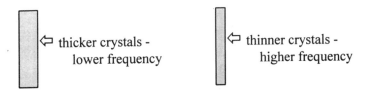

⇦ thicker crystals -   ⇦ thinner crystals -
   lower frequency   higher frequency

The typical propagation speed for piezoelectric material is 4-6 mm/μs, about 3 to 4 times faster than that for soft tissue.

RESONANT FREQUENCY is the "operating frequency" of a transducer. A pulsed wave PZT crystal oscillates its resonant frequency. It is also called the natural frequency.

$$\text{frequency (MHz)} = \frac{\text{material's propagation speed (mm / us)}}{2 \text{ x thickness (mm)}}$$

Frequency is *inversely related to crystal thickness:*
- The thinner the crystal, the higher the frequency.
- The thicker the crystal, the lower the frequency.

Frequency is *directly related to the speed of sound in the crystal:*
- The higher the speed, the higher the frequency.
- The lower the speed, the lower the frequency.

> **For Pulsed Transducers:**
> The thinner the active element, the higher the transducer's frequency. (Think of a crystal glass.)
> The faster the active material's propagation speed, the higher the transducer's frequency.

Note: The thickness of the PZT is one-half the wavelength of sound traveling in the crystal.

Transducers that emit high-frequency sound are made with very thin PZT crystals with very fast propagation speeds.

The lowest frequency transducers have thick PZT crystals with slow propagation speeds.

BANDWIDTH

Because of the effects of the damping material, an imaging transducer doesn't emit a sound pulse composed of only a pure, single frequency. Rather, the pulse contains *a range of frequencies* (the bandwidth) below and above the main, or "resonant," frequency. The **BANDWIDTH** is the difference between the highest and the lowest frequency found in the pulse.

Units: Hz

The narrower the bandwidth, the more exact the frequency emitted by the transducer. The damping material tends to increase the bandwidth of a transducer.

The process of damping increases the range of frequencies present in a sound pulse. The shorter the pulse, the wider the bandwidth. Recall that short pulses are required to create high-quality images.

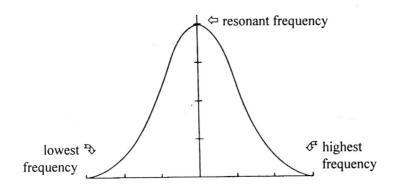

Thus, a "5 MHz" transducer actually emits a pulse composed of multiple frequencies, for example, ranging from 3.5 to 6.5 MHz. The center frequency (also called resonant or main) is 5 MHz. In this case, the bandwidth is 3 MHz (6.5 - 3.5 = 3). Frequencies above and below the resonant frequency are contained in the pulse.

Imaging transducers tend to have a wider bandwidth than transducers used in therapeutic ultrasound.

QUALITY FACTOR

The **QUALITY FACTOR** (Q-factor) is a unitless number that represents the ability of a transducer to emit a "clean" ultrasound pulsed with a narrow bandwidth. Imaging transducers tend to have lower Q-factors (in the range of 2 - 4) than ultrasound transducers used in therapeutics because imaging transducers create short pulses with wide bandwidths. Short pulses contain sound with a broad range of frequencies and have a low Q-factor. Short pulses are required to make high-quality images.

$$\text{Quality Factor} = \frac{\text{resonant frequency (MHz)}}{\text{bandwidth (MHz)}}$$

QUESTIONS - TRANSDUCER MATERIAL

Are the following statements true or false?

1. If the frequency of the electrical excitation voltage of a pulsed wave *F* transducer is 6 MHz, then the sound wave's frequency is 6 MHz.

2. If the pulse repetition frequency of a transducer is increased, then the frequency of US produced by the transducer remains the same.

3. The diameter of the active element of a transducer helps to determine the *F* frequency of the US produced by the transducer.

4. If the frequency of the electrical excitation voltage of a continuous wave *F* transducer is 6 MHz, then the frequency of the acoustic wave is 6 MHz.

5. Two piezoelectric crystals are made from the same material (Hint: The *F* propagation speeds must be the same.) The thicker crystal will make a pulsed transducer with a higher resonant frequency.

6. Two active elements are made from the same PZT. The thicker crystal will *F* make a continuous wave ultrasound transducer with a lower frequency.

7. The normal propagation speed for piezoelectric material is 10 to 20 times *F* greater than that for soft tissue.

8. The acoustic impedance of the matching layer is equal to the acoustic *F* impedance of skin.

9. Imaging transducers typically have a narrower bandwidth than do *F* therapeutic ultrasound transducers.

10. Imaging transducers tend to have lower quality factors than those of transducers used in therapeutic applications.

True or False? The damping material in a transducer acts to:
  11. increase the sensitivity.   12. increase the pulse length.
  13. decrease the pulse duration. 14. improve the system's image quality.
  15. decrease the bandwidth.   16. decrease the quality factor.

Answers: 1- F.   2- T.   3- F.   4- T.   5- F.   6- F.   7- F.   8- F.
             9- F. 10- T. 11- F. 12- F. 13- T. 14- T. 15- F. 16- T.

QUESTIONS - TRANSDUCERS

1. Which of the following crystals will produce a sound pulse with the lowest frequency?
   a) Thin PZT crystal with a low propagation speed.
   b) Thin PZT crystal with a high propagation speed.
   c) Thick PZT crystal with a slow propagation speed.
   d) Thick PZT crystal with a fast propagation speed.

2. A pulsed ultrasound transducer has a resonant frequency of 5 MHz. The lowest frequency found in the pulse is 2 MHz, and the highest frequency is 8 MHz. What is the bandwidth of the transducer?
   a) 5 MHz   b) 8 MHz   c) 3 MHz   d) 6 MHz   e) 2 MHz

3. For the transducer described in question 2, what is the resonant frequency?
   a) 5 MHz   b) 8 MHz   c) 3 MHz   d) 6 MHz   e) 2 MHz

4. For the transducer described in question 2, what is the quality factor?
   a) 0.83   b) 8 MHz   c) 1.2   d) 5 MHz   e) 5

5. Which type of transducer has a greater Q-factor -- therapeutic or imaging?

6. Which type of transducer has a greater bandwidth -- therapeutic or imaging?

7. Which type of transducer has more backing material -- therapeutic or imaging?

8. In an imaging transducer, the purpose for attaching the backing material to the PZT is to:
   a) increase the bandwidth.
   b) decrease the Q-factor.
   c) improve image quality.
   d) decrease the transducer's sensitivity.
   e) increase the frequency.

Answers: 1 - c.   2 - d.   3 - a.   4 - a.   5 - therapeutic.
                6 - imaging.   7 - imaging.   8 - c.

SOUND BEAMS

A sound beam is shaped like an hourglass. As sound travels, the width of the
beam changes. At its starting point, it is exactly the same size as
the diameter of the transducer. The beam gets progressively
narrower until it reaches its smallest diameter. Then, it diverges.

Focus - the **location** where the diameter of the sound beam is at a minimum.

Focal length - (near zone length) the **distance** from the transducer face to the
location where the beam reaches its smallest diameter.

Near zone (FRESNEL ZONE or near field) - the **region** between the
transducer and the focus.

Far zone (FRAUNHOFER ZONE or
far field) - the **region**
deeper than the near field.

> Near zone: Short name - Fresnel.
> Far zone: Long name - Fraunhofer.

Focal zone - the region surrounding the focus that generally has a narrow
beam. The image quality is best at the depths where the beam is
the narrowest.

Anatomy of a sound beam:

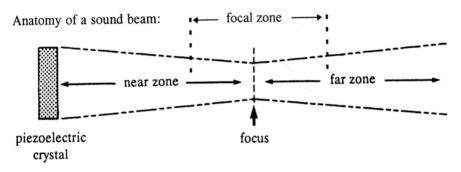

piezoelectric
crystal

focus

The focal length (or the depth of the beam's waist) is determined by two
factors:
  • The diameter of the PZT crystal.
  • The emitted frequency.

> Larger crystal diameter ⇨ longer focal length.
> Higher emitted frequency ⇨ longer focal length.

The focal depth is directly related to crystal diameter:
- Large-diameter crystals produce a beam with a deep focus.
- Small-diameter crystals produce beams with a shallower focus.

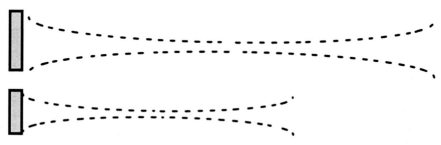

Additionally, the focal depth is directly related to the frequency of the sound:
- High-frequency waves have a deep focus.
- Low frequency waves have a shallow focus.

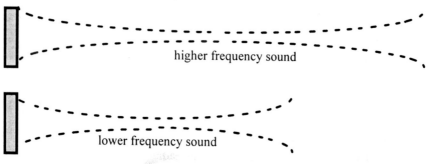

higher frequency sound

lower frequency sound

Note: The fact that high-frequency sound creates a deeper focus is particularly troublesome to sonographers. Sonographers often think "How can this be? High-frequency beams must have a shallow focus, because high frequencies cannot penetrate to substantial depths." The explanation is: Higher frequency sound creates a beam with a deeper focus. However, this limits the clinical utility of high-frequency probes. The transducer manufacturer is aware of the conflict and applies the following solution. The manufacturer also knows that when the crystal has a small diameter, the focus is shallow. Therefore, in order to create a high-frequency transducer with a shallow focus, a high-frequency crystal is fabricated with an extremely small diameter. The crystal's small diameter tends to create a shallow focus. Thus, the tendency to make a deep-focus beam with high frequencies is overcome by the crystal's very small diameter. This combination creates a clinically useful transducer.

Unfocused disc transducer operating in the continuous mode: At the end of the
near zone, the beam diameter is one half the transducer diameter. At
two near zone lengths, the beam diameter equals the transducer
diameter.

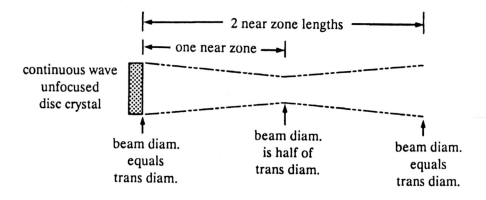

DIFFRACTION AND HUYGEN'S PRINCIPLE

When sound waves are produced by a small sound source -- approximately the
size of the wavelength of the sound -- the wave will diverge as it
propagates. This divergence, or spreading, is called
DIFFRACTION. Pulses produced by imaging transducers do not
display a diffraction pattern because the PZT crystal, acting as the
source of the sound wave, is considered large rather than small.

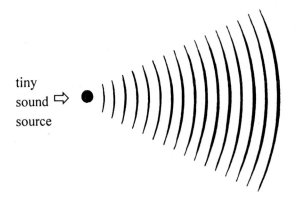

HUYGEN'S PRINCIPLE - Sound waves produced by imaging transducers do not diffract because they obey Huygen's principle. Imagine that the active element is composed of an infinite number of small sound sources. Each particle on the surface of the PZT can be considered a small, individual sound source that produces a little sound wave. The numerous, tiny diffracted wave patterns interfere with each other. The sound beam resulting from the destructive and constructive interference of the sound wavelets is hourglass-shaped. (See page 14 for a discussion of wave interference.)

According to Huygen's Principle, when all of these wavelets are combined, they produce a sound beam with most of the energy transmitted along a main, central beam having the shape of an hourglass.

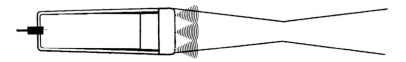

With large sound sources, tiny wavelets interfere to produce a sound beam.

SOUND BEAM DIVERGENCE

As sound propagates into the far field, the beam spreads out or "diverges."
DIVERGENCE is the extent to which the beam flares in the far
field.

Beam Divergence - In the far zone, small-diameter crystals produce more
divergent beams. Therefore, in the far zone, small crystal
diameters will result in wider US beam diameters.

As the transducer diameter increases, the beam is less divergent.

As the transducer diameter decreases, the beam diameter in the far field
increases. Smaller diameter crystals produce sound beams that
diverge dramatically in the far field.

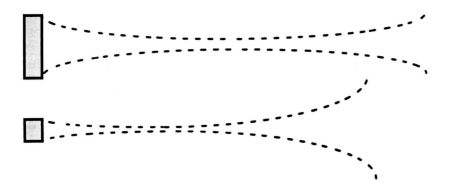

PZT CRYSTALS AND BEAM CHARACTERISTICS

Recall that specific attributes of the PZT crystal affect the characteristics of the
sound beam:

Beam Characteristic	Crystal Attributes
• frequency of pulsed transducers	thickness of PZT propagation speed of PZT
• focal depth	crystal diameter frequency (thickness & speed of crystal)
• beam divergence	crystal diameter

REVIEW - SOUND BEAMS

1. Which of the following transducers has the shallowest focus?
    a) 4 mm crystal diameter, 5 MHz   b) 8 mm crystal diameter, 3 MHz
    c) 4 mm crystal diameter, 3 MHz   d) 8 mm crystal diameter, 5 MHz

2. Which transducer in question #1 has the deepest focus?

3. Which transducer in question #1 is most likely to be clinically successful in
    producing high-quality images of superficial structures in the
    body?

4. In what location is a sound beam's diameter the narrowest?

5. The distance from the transducer to the minimum beam diameter is called
    the _____.

6. What is another name for the near zone?

7. What is another name for the far zone?

8. What is the shape of a sound beam produced by a small sound source?

9. What is the shape of a sound beam produced by a large sound source, such
    as a diagnostic imaging transducer?

10. The region where a sound beam gets progressively narrower is the

    _____.

11. The region where a sound beam gets progressively broader is the

    _____.

12. Which of the following transducers creates a sound beam that will
    diverge the least in the far zone?
        a) 5 mm crystal thickness   b) 10 mm crystal diameter
        c) 10 mm crystal thickness   d) 5 mm crystal diameter

ANSWERS - SOUND BEAMS

1 - c: The crystal with the lowest frequency and smallest diameter will
           create the shallowest focus.

2 - d: The crystal with the highest frequency and the largest diameter will
           create the deepest focus.

3 - a: High-quality superficial images are obtained by use of a high-
           frequency sound beam with a shallow focus. This compromise is
           best achieved with the transducer described in choice a. Note that
           the transducer described in choice c has a shallower focus, but its
           lower frequency creates a lower quality image.

4 - Focus.

5 - near zone length or focal length

6 - Fresnel zone.

7 - Fraunhofer zone.

8 - A diffraction pattern, appearing like an ever-expanding wake produced by
        a boat moving along the surface of a lake.

9 - An hourglass shape.

10 - A sound beam gets progressively narrower in the near zone.

11   A sound beam diverges or flares out in the far zone.

12 - b: Sound beams produced by large-diameter crystals diverge the least in
           the far zone.

FOCUSING

FOCUSING causes the waist, or middle, of the sound beam to get narrower. This narrowing creates a more accurate image.

Focusing is mainly effective in the near field and the focal zone.

Four methods of focusing:

+ Lens:

+ Curved Piezoelectric Crystal (also called Internal Focusing):

+ Focusing Mirror:

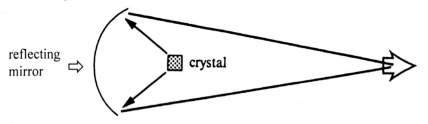

+ Electronic Focusing - used in phased array technology (See page 107).

> Focusing with a lens, curved crystal, or mirror is called "conventional."
> Electronic focusing is used in phased array technology.

FOCUSING AND SOUND BEAM ANATOMY

FOCUSING narrows the waist of an ultrasound beam. It improves image
quality.

When a beam is strongly focused, the waist becomes very narrow.
This creates an excellent image in the focal zone.

Two additional effects of focusing are:
- a shorter focal zone. This decreases image quality in regions
  other than the focal zone.
- a shallower focal depth. The focus moves toward the
  transducer. In other words, the near zone length is shorter.

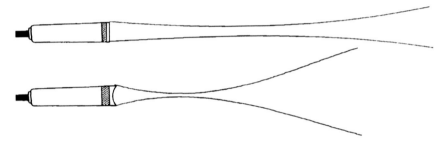

RESOLUTION

As previously discussed, specific factors associated with sound pulses
determine the ability to produce high-quality images. The term
resolution applies to image quality.

RESOLUTION describes the ability to accurately image structures.

Two characteristics of a pulse that improve
image quality:
- Shorter pulses.
- Narrower pulses.

Shorter pulses more accurately identify
structures that lie close together when
one is in front of the other. This is
called **longitudinal resolution**.

The term **spatial resolution**
is used to describe an US
system's overall ability to
accurately create images of
small structures in their
correct anatomic position.

Narrow pulses provide the ability to accurately identify structure that lie close
together side-by-side. This is called **lateral resolution**.

LONGITUDINAL RESOLUTION

Longitudinal resolution - the ability to distinguish two structures that are close to each other FRONT to BACK.

> **L**ongitudinal
> **A**xial
> **R**ange, **R**adial
> **D**epth

> Use the acronym **LARD** to remember all five names for AXIAL resolution.

Units: mm, or any unit of
distance. The smaller the number, the better the picture quality.

The numerical value for the LARD resolution answers the following question:
How close can two structures be *(with one in front of the other)* and still produce two distinct echoes on an ultrasound image?

LARD resolution is determined by the spatial pulse length (see p. 32), with shorter pulses creating better images.

$$\text{LARD resolution (mm)} = \frac{\text{spatial pulse length (mm)}}{2}$$

Note: As frequency increases, the numerical value of the LARD resolution decreases. This means that high-frequency transducers produce pulses with better LARD resolution and higher quality images.

Additionally, transducer pulses are designed with a minimum number (2-4) of cycles in each pulse. This means that the numerical value of LARD resolution is low, and the image is of exceptional quality.

LARD resolution is determined by the sound source and the medium (because it is dependent upon spatial pulse length). In diagnostic imaging, LARD resolution typically ranges from 0.05 to 0.5 mm.

REVERBERATIONS

REVERBERATIONS are multiple, equally spaced echoes or reflections that may occur when two strong reflectors lie in the line of the ultrasound beam.

The echoes that are created as the US "ping-pongs" back and forth between the two reflectors may create images that do not represent reality.

Reverberation is a type of ARTIFACT or an error in imaging (see page 202), which often has the appearance of rungs on a ladder.

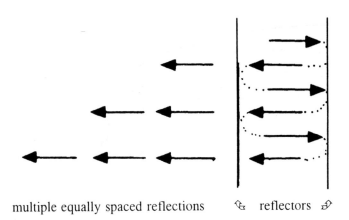

multiple equally spaced reflections   ☜ reflectors ☞

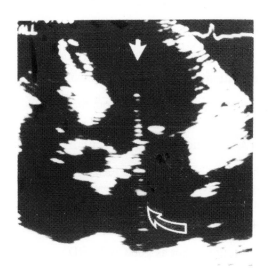

The horizontal linear echoes identified by the white arrows are **reverberations** from a structure within the cardiac chamber. The true location of the structure is marked by the solid white arrow. (Source: Leonard Pechacek, Houston, TX)

LATERAL RESOLUTION

Lateral Resolution - **L**ateral
Angular
Transverse
Azimuthal

> Use the acronym "**LATA**-rel" to remember the four different names for lateral resolution.

LATERAL (LATA) RESOLUTION is
the minimum distance that two side-by-side structures can be separated and still produce two distinct echoes.

The numerical value for the LATA resolution answers the following question: What is the minimum distance that two structures, positioned *side by side,* can be apart and still produce two distinct echoes on an ultrasound image?

Units: mm, any unit of length.

The lateral resolution is approximately equal to the **BEAM DIAMETER**.

Since the beam diameter *varies with depth*, the lateral resolution also varies with depth.

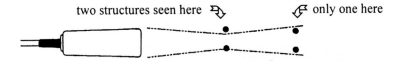

two structures seen here ➘   ➘ only one here

To determine whether two structures will produce two distinct echoes on an image, compare their separation with the beam diameter.

The lateral resolution is best at one near zone (focal) length from the transducer, since the sound beam is narrowest at that point. Focusing improves LATA resolution in the focal zone.

Note: Lateral resolution varies with depth. It is best at the focus.

> As the lateral resolution becomes numerically smaller, the picture improves visually.

Note: In clinical imaging, LATA resolution is not as good as LARD resolution. A sound pulse is typically wider than its length.

QUESTIONS - SOUND BEAMS AND LATA RESOLUTION

The near zone length of a transducer system is 8 cm. The beam is unfocused.
The transducer diameter is 9 mm.

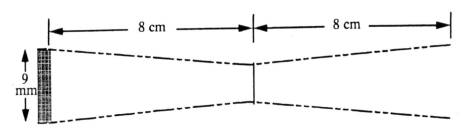

1. What is the lateral resolution at a depth of 8 cm?

2. What is the lateral resolution at a depth of 16 cm?

3. What is the best estimate for the resolution at 4 cm?

   a) 1 cm   b) 9 mm   c) 5 mm   d) 14 mm

4. What is the best estimate for the resolution at 7 cm?

   a) 1 cm   b) 9 mm   c) 5 mm   d) 14 mm

5. What is the best estimate for the resolution at 21 cm?

   a) 1 cm   b) 9 mm   c) 5 mm   d) 14 mm

The frequency of a transducer does not change. If the diameter of the new
piezoelectric crystal increases, what happens to the:

   6) near zone length?   7) beam diameter in the far zone?

   8) wavelength?   9) beam diameter in the near zone?

Answers: 1 - 4.5 mm.   2 - 9 mm.   3 - c.
                  4 - c.   5 - d.   6 - It increases.
                  7 - It decreases. 8 - No change.   9 - It increases.

REVIEW - RESOLUTION

1. The ability to distinguish two structures lying close together is called

    _____.

2. The ability to distinguish two structures lying close together front-to-back is
    called _____.

3. The ability to distinguish two structures lying close together side-by-side is
    called _____.

4. Axial resolution and lateral resolution are both measured with units of

    _____ .

5. When the number of cycles in a pulse increases while the frequency
    remains the same, the numerical value of the range resolution is
    _____ (greater, lesser, the same).

6. When the number of cycles in a pulse increases while the frequency
    remains the same, the image quality _____ (improves,
    degrades, remains the same).

7. _____ (high, low) frequency transducers have the best range resolution.

8. Name the four synonyms for longitudinal resolution.

9. Name the three synonyms for lateral resolution.

10. The length of a pulse is 8 mm. What is minimum distance between two
    reflectors, positioned one in front of the other, that still produces
    two echoes on our image?
    a) 8 mm   b) 4 mm   c) 16 mm
    d) 2 mm   e) cannot be determined

ANSWERS - RESOLUTION

1 -   spatial resolution

2 -   longitudinal, axial, range, radial, or depth resolution

3 -   lateral, angular, transverse or azimuthal

4 -   distance, such as mm

5 -   The more cycles there are in a pulse, the longer the pulse becomes. The numerical value of the range resolution *increases*

6 -   When the number of cycles increases, the spatial pulse length increases and the image quality *degrades*

7 -   High frequency

8 -   longitudinal, axial, range, radial and depth (acronym: LARD)

9 -   lateral, angular, transverse, and angular (acronym: LATA)

10 - b) 4 mm: This value is one-half of the pulse's length.

DISPLAY MODES

Four methods of displaying ultrasonic echoes are discussed below.

A - MODE - amplitude mode

When an ultrasound pulse is emitted by the transducer, a dot moves at
constant speed across the screen of the ultrasound system. When
an echo returns, an upward deflection, proportional to the
amplitude of the returning echo, is placed on the screen.

The image looks like a "Manhattan" skyline.
The height of deflection relates to the amplitude of returning echo.
The location of the deflection relates to the reflecting structure's depth.

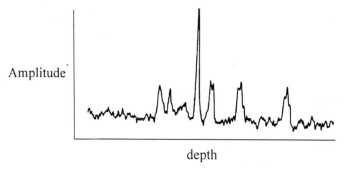

depth

B - MODE - brightness mode

When an US pulse is emitted by the transducer, an invisible dot moves at a
constant speed across the screen of the ultrasound system. When
an echo returns, the dot lights up with a brightness that is
proportional to the amplitude of the returning echo.
Returning echoes appear as spots on the line of travel of the
emitted US pulse. The stronger the returning echo, the brighter
the spot.

The gray-shade of the dot relates to the strength of returning echo.
The location of dot relates to the depth of the reflecting structure.

M - MODE - motion mode

Dragging a photosensitive paper across a B-mode display causes the production of **squiggly lines** on the paper. These lines represent the position and motion of the reflecting surfaces as they occur in time. M-mode is the only mode that provides information on the position of reflectors with respect to time.

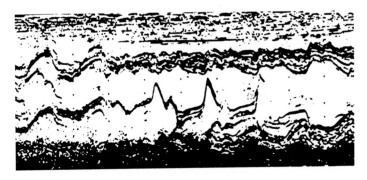

C - MODE - constant depth mode

By using gating electronics, only echoes returning from a specific depth in the body are processed. The sonographer selects this depth. This special form of B-scanning produces images from a specific plane or slice that is located at a particular depth in the body.

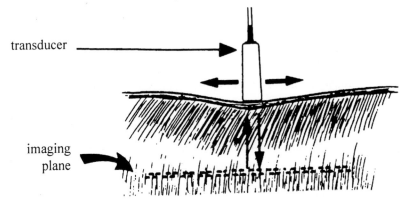

TWO-DIMENSIONAL IMAGING

Generally, the four modes previously discussed provide echo data in a one-
dimensional format. However, we desire "images" and two-
dimensional slices of anatomy to diagnose our patients.

Two factors make it difficult to create a two-dimensional image:
- Ultrasound only travels in a straight line.
- Narrow beam widths are required for high-quality images with
good lateral resolution.

To overcome these difficulties, a two-dimensional image is constructed from
multiple US pulses.

B - SCAN

B-mode - provides data along a straight line as an US beam propagates.

If the location and angulation of the transducer are known by the ultrasound
system, it is possible to create a two-dimensional image from
multiple B-mode pulses.

The transducer is attached to the ultrasound system by using an articulated arm
(similar to the structure of a dentist's drill). At each joint in the
armature, a sensor measures the position of the transducer and the
direction that the US beam is directed. These sensors are called
position discriminators or **potentiometers**.

Each line of the image is a result of a single ultrasound pulse.

The sonographer moves the transducer through an appropriate path while the
data are collected and added together over time to form a
composite picture.

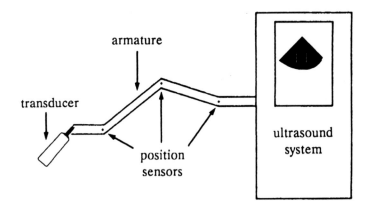

B- Scanner

ADVANTAGES	DISADVANTAGES
• Large images.	• Long time to complete scan.
• Good resolution.	• Patient movement destroys image.
	• Cannot image moving structures.
	• Requires large acoustic window.

REAL-TIME IMAGING

REAL-TIME IMAGING is the production of "motion pictures" of the anatomy. This type of imaging consists of a series of frames or pictures displayed in a rapid sequence, which give the impression of constant motion.

Motivation for real time scanning:
- Overcome the shortcomings of B-scanning.
- Image moving structures.
- Perform Doppler ultrasound.

Goal: To produce pictures or frames as rapidly as possible, in order to faithfully image moving structures.

Solution: Have the US system create an imaging plane by accurately and automatically steering the beam through a predetermined pathway.

Acquire a large number of scan lines.

Combine the scan lines into a two-dimensional picture.

TEMPORAL RESOLUTION is resolution pertaining to time and motion. It defines the ability to accurately locate the position of moving structures at particular instants in time. The greater the number of frames created per second, the better the temporal resolution.

FRAME RATE is the number of frames or images created each second. It is reported in **units** of per second or Hertz.

PULSES, LINES, AND FRAMES

Each frame of an ultrasound image is made of individual scan lines.

On the display, a single scan line (B-mode) is created by a pulse from the transducer. Think of each scan line as a spoke of a bicycle wheel.

The transducer's beam is automatically steered to a slightly different direction and then fired again. This process is repeated numerous times.

If there are 100 lines in an image, then 100 pulses are required to make the picture. One pulse of ultrasound is needed for each scan line.

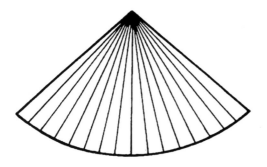

MECHANICAL SCANNING

A **MECHANICAL TRANSDUCER** contains a single, disc-shaped active element that is physically moved or rotated through a pathway. Thus, the ultrasound beam is steered mechanically.

It produces a fan- or sector-shaped image.

Curvature of the active element, an acoustic lens or a reflecting mirror, focuses the beam at a specific depth. This is called **conventional focusing**.

Focusing occurs in both horizontal vertical planes.

Mechanical scanners are fixed-focus. Changing the focus requires changing the scanhead.

(Source: Toshiba)

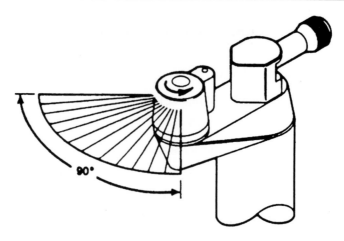

(Source: Advanced Technology Laboratories)

Mechanical scanheads have the advantage of having contact with a small area of the skin, called a small **acoustic footprint**. Thus, mechanical scanheads are successful for imaging though small "acoustic windows," such as between ribs. Additionally, they are relatively simple and inexpensive to manufacture.

However, a disadvantage of using mechanical scanners is that as the beams reach ever increasing depths, they get further apart. (Visualize the spokes of a bicycle wheel. The further away from the hub, the more space between them.) This phenomenon results in decreased spatial resolution and image quality at increased depths.

Additionally, mechanical transducers may create side lobes:

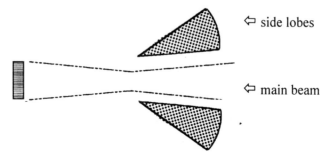

⇦ side lobes

⇦ main beam

Note: When the piezoelectric crystal (PZT) of a mechanical scan head malfunctions, the entire image vanishes.

TRANSDUCER ARRAYS

An **ARRAY** is a collection of active elements within a single transducer
housing.

A single slab of piezoelectric material is cut into a collection of separate pieces
called **elements**.

Each active element is connected to its own electronic circuitry. Each element
is isolated from its neighboring crystals, both electronically and
acoustically.

Linear array - a collection of elements arranged in a line. There are four
types:

- ◆ Linear switched or sequential array.

- ◆ Linear phased array.

- ◆ Convex, curved, or curvilinear sequential array.

- ◆ Convex, curved, or curvilinear phased array.

Annular array - a collection of ring-shaped elements with a common
center.

LINEAR SWITCHED ARRAYS

LINEAR SWITCHED ARRAYS are large transducers (2 - 8 cm long) with
multiple elements arranged in a line; they are also called linear
sequential or linear arrays.

Elements are fired in a sequence to create a
two-dimensional image that
consists of parallel scan lines
emitted at different points along
the face of the transducer.

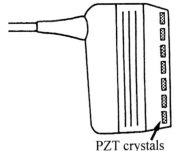

PZT crystals

Linear switched arrays have the following characteristics:

- Rectangular image shape.

- Conventional focusing - internal, lens, or mirror.

- No beam steering - The sound beams simply are directed outward from each element.

(Source: Toshiba)

Advantages of linear switched arrays:

- The beams creating the image are parallel at all depths.

- There are no moving parts to wear out.

Disadvantages of linear switched arrays:

- A large contact area with the skin, called an acoustic footprint, is required to construct an image.

- They cannot be used when the acoustic window is small.

When a single piezoelectric (PZT) crystal of a linear array malfunctions, a single line of image data extending downward from the transducer drops out. The remainder of the image is essentially unaffected.

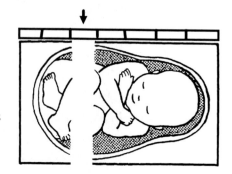

CONVEX SWITCHED ARRAYS

CONVEX SWITCHED ARRAYS have a collection of active elements
arranged in an **arc**; they are also called convex sequential arrays
or curvilinear arrays. Since the crystals are arranged in an arc,
the imaging plane has a natural sector shape.

Elements are fired in a sequence
to produce a two-
dimensional image. elements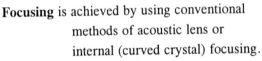
The image is
created by scan
lines radiated at
different points by different elements along the face of the probe.

The image is blunted **fan-** or **sector-shaped**.
The superficial region of the
image does not reach a point;
rather it has a broad, curved
shape.
Focusing is achieved by using conventional
methods of acoustic lens or
internal (curved crystal) focusing.

There is **no beam steering** because the curved architecture, by its very design,
creates a sector.

(Source: Acuson)

Advantages of convex switched arrays:

  ◆ Natural sector image.
  ◆ No moving parts.

Disadvantages of convex switched arrays:

  ◆ They require a large acoustic footprint.
  ◆ They are difficult to use when the acoustic window is small.
  ◆ Sound beams tend to separate from each other, leaving gaps, as the imaging depth increases.

When a single piezoelectric (PZT) crystal of a convex switched array malfunctions, a line of image data extending downward from that crystal drops out. The remainder of the image is essentially unaffected.

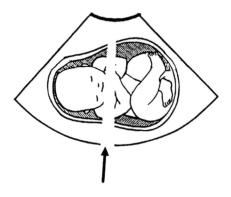

LINEAR PHASED ARRAYS

In **LINEAR PHASED ARRAYS**, the piezoelectric (PZT) elements are arranged in a straight line, but the array is very small.

The image is *fan- or sector-shaped.*

A collection of electrical spikes in various patterns is delivered to all of the elements. These patterns serve to focus and steer the US beam.

Multiple electronic signals are used to create a single acoustic pulse. All of the crystals in the phased array transducer are fired nearly simultaneously, each crystal by its own electronic spike. Each of the PZT crystals in the array produces a small sound wave. These multiple waves interfere with each other, creating a single sound pulse that is focused and steered. See page 14 for a description of constructive and destructive interference.

There are minuscule (microsecond) time delays between electronic pulses delivered to the array elements.

Beam steering is electronic. Sector angles of up to 90° are created while all components of the transducer remain motionless.

Focusing is electronic. Since focusing is achieved via the ultrasound system's electronics, the focal depth on phased array systems is *controlled by the sonographer.*

(Source: Hewlett-Packard)

The US system alters the electronic pattern that excites the crystals. The pattern is changed for each successive sound pulse, so that the beam can be swept to create an imaging plane.

Similarly, time delays during reception are applied to the electrical signals returning from the transducer to the ultrasound system. This is called **dynamic receive focusing** or **dynamic focusing**. Receive zone focusing can selectively process returning echoes and optimize image quality.

| Electronic curvature | ⇨ | beam focusing |
| Electronic slope | ⇨ | beam steering |

Advantages of linear phased arrays:

- ◆ No moving parts.
- ◆ Sonographer controlled (dynamic) focusing.
- ◆ Small acoustic footprint.

Importance of dynamic focusing:

Conventional focusing (lens, internal or mirror) creates a sound beam with a particular fixed focal length. With phased arrays, the electronic delay pattern determines the depth of the focus. Since the pattern can be changed, so too can the location of the focus. Thus, a sonographer may adjust the controls of the ultrasound system to manipulate the depth of the focus to correspond to the depth of the clinically important anatomy!

Note that a single sound pulse has its individual focus at a particular depth. However, multiple pulses, each with a progressively deeper focus, may be directed to follow a single scan line on the image. The image data from each focal zone is stored, while the remainder of the information is discarded. A single line of image data is a combination of focal regions from a number of sound pulses. This provides superb image quality. Unfortunately, using this technique to create images requires substantial time and, thus, decreases frame rate.

Disadvantages of linear phased arrays:

- ◆ Expense and complexity.
- ◆ Potential for grating lobe artifacts.

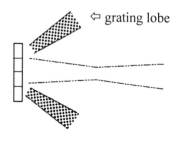

⇦ grating lobe

When a single crystal in a phased array transducer malfunctions, the steering and focusing of the sound beam become erratic. This may express itself as a minor problem, or it may be so severe that the clinical utility of the transducer is destroyed.

The time delays may be thought to represent the surface of a reflecting mirror; the direction and the focusing of the beam become apparent.

Note: To learn how the characteristics of the electrical pattern determine beam steering and focusing, study this page and the next.

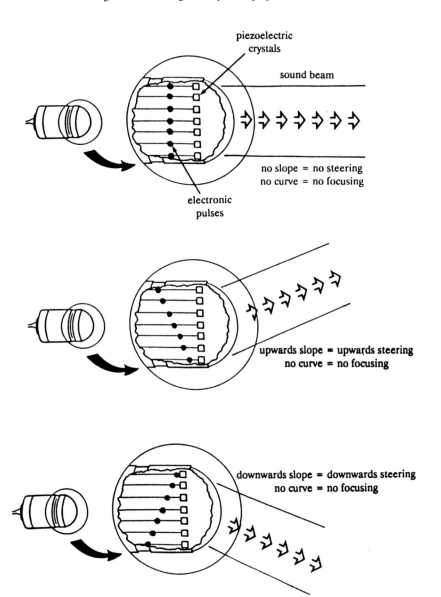

piezoelectric crystals

sound beam

no slope = no steering
no curve = no focusing

electronic pulses

upwards slope = upwards steering
no curve = no focusing

downwards slope = downwards steering
no curve = no focusing

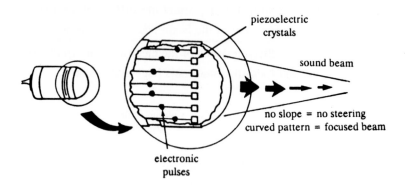

piezoelectric
crystals

sound beam

no slope = no steering
curved pattern = focused beam

electronic
pulses

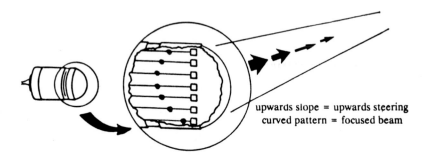

upwards slope = upwards steering
curved pattern = focused beam

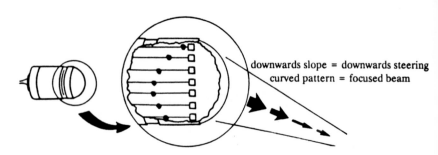

downwards slope = downwards steering
curved pattern = focused beam

ANNULAR PHASED ARRAYS

ANNULAR ARRAY TRANSDUCERS have concentric
rings (donut-shaped) cut from the same circular
slab of piezoelectric material.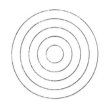

Steering is performed **mechanically** and creates a sector-
shaped image.

With respect to focusing, recall the following:
- Small diameter rings have a shallow focus but diverge rapidly.
- Large diameter rings have a deep focus.

Focusing strategy - selected focal zones. Annular arrays use inner crystals for
shallow regions and outer crystals for deep regions. Multiple
sound pulses are used for each scan line on the image.

Phasing provides **electronic focusing** in all planes at all depths; a core sample.

Annular phased array beam profile

ringed
elements ▶

Identical dynamic extended focus in vertical and horizontal
planes for tight focus and thin tomographic slice

(Source: Advanced Technology Laboratories)

Advantages of annular phased array:

- ◆ Multiple foci -- superior image quality at all depths.
- ◆ Small acoustic footprint -- does not require large contact area with skin.
- ◆ Small window imaging -- can image between ribs.

Disadvantages of annular phased array:

- ◆ Multi focus -- requires a long time to make a single image.
- ◆ Low frame rates -- reduced temporal resolution.
- ◆ Mechanical steering -- transducers have moving parts.

If an element in an annular array malfunctions, there is a band of image drop-out at a particular depth. The regions above and below the drop-out are unaffected by the malfunction .

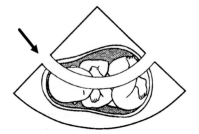

CONVEX PHASED ARRAYS

Convex phased arrays consist of multiple PZT crystals arranged in a **curved architecture.** The array is small. This curved architecture helps to create the **sector-shaped image**.

Similar to a linear phased array, steering and focusing are achieved by electronic patterns.

When a single crystal in a convex phased array transducer malfunctions, the steering and focusing of the sound beam become erratic. This may appear on the image as a minor problem, or it may be so severe that the clinical utility of the transducer is destroyed.

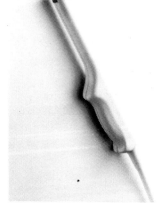

(Source: Advanced Technology Labs)

Advantages and disadvantages are similar to those of linear phased array transducers.

COMBINED TRANSDUCER DESIGNS

Each type of transducer previously described has specific strengths and weaknesses. To optimize the image quality and clinical utility of their systems, equipment manufacturers have combined different transducer technologies into a single probe. For example, phased electronics have been introduced into linear sequential array probes to provide electronic steering and multiple focal zones. This combination is called a **VECTOR ARRAY**.

The subject of combined technology transducers is worthy of an entire textbook. The practicing sonographer should appreciate that the best transducer is one that provides *optimal image quality while meeting the challenges of the specific clinical setting*. By combining several fundamental transducer designs into a single probe, the manufacturers have attempted to meet this challenge.

The following guidelines should aid in identifying the technological basis for any particular transducer:

- If a transducer has an adjustable focus, focusing is achieved by using phased array technology.

- If a transducer has multi focus capability, focusing is achieved by using phased array technology.

- If the transducer has no moving parts, there is either an absence of steering (sequential arrays) or there is electronic steering (phased arrays).

- If the sonographer can "angle" the image produced by a linear sequential array and change the shape of the image from a rectangle into a trapezoid, steering is achieved electronically by using phased array technology. This is a **vector array**.

VECTOR ARRAYS

VECTOR ARRAYS combine linear phased array and linear sequential array
technologies.

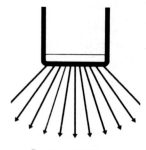

Source: Acuson

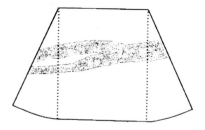

Source: Hewlett-Packard

The image is trapezoidal in shape. The image is blunted **fan-** or **sector-shaped**.
The superficial region of the image does not reach a point; rather
it originates from a flat-top.

WATER PATH SCANNERS

WATER PATH SCANNERS have a water bath, acoustic standoff, or offset
built into the probe.

The bath is placed between the sweeping or steering mechanism of the
transducer and the patient.

A large transducer face or "acoustic footprint" is placed on the patient.

Superficial structures are imaged more clearly with water path scanners. This
is the rationale for **small-parts scanners**.

They allow contact between transducer and patient when direct contact
scanning is difficult.

TRANSDUCERS AND IMAGE SHAPE

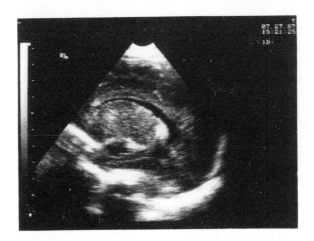

A mechanical scanner produced this image of a neonatal brain. Note the sector
shape of the image. Steering is achieved by moving the active element
mechanically, whereas focusing is achieved by use of a lens or an
internal curved crystal. (Source: Corometrics Medical Systems)

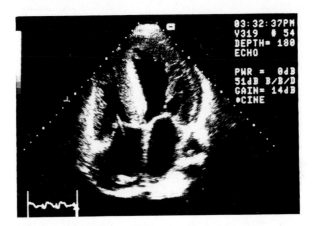

An image of the heart produced with a phased array system. Note that the
sector-shaped image is similar to that of a mechanical scanner. Both
beam steering and focusing are achieved electronically, with tiny time
delays in the pulses delivered to the array elements. (Source: Acuson)

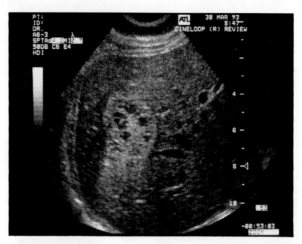

A scan of the liver, displaying a mass. This scan was produced with an
     annular phased array transducer. Note that the sector shape of the
     image is identical to the mechanical and phased array systems. The
     beam is steered by a mechanical method. However, the beam is
     focused with an electronic, phased array approach.
     (Source: Advanced Technology Laboratories)

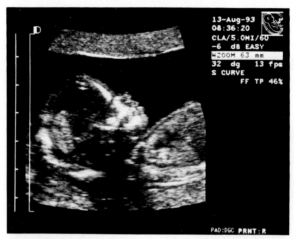

A scan of a fetal profile. The image was produced by a curved array
     transducer. Note that a sector image is obtained, but this sector differs
     from the previous three examples. The region near the transducer is
     much wider than the width shown in the other examples because of the
     large size of the transducer. The sector shape is produced without
     steering because of the curvature of the active elements in the array.
     Focusing is achieved by conventional methods (lens or curved
     crystals). (Source: Diasonics, Inc.)

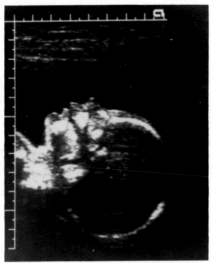

A scan of a fetal profile obtained with a linear switched or sequential array transducer. Note the rectangular image shape. The width of the image is determined by the width of the crystal array. With pure linear switched arrays, there is no beam steering, and focusing is achieved by conventional means (lens or curved crystal). (Source: Acuson)

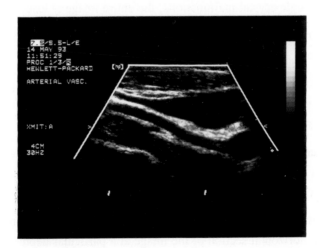

A scan of the vertebral arteries obtained with a vector transducer. Note the trapezoidal image shape. This probe combines linear phased array and linear switched array technologies. The region near the transducer is much wider than that shown in the other sector examples because of the large transducer size. The top of the scan is flat because the transducer is similar to a linear switched array. (Source: Hewlett-Packard)

TRANSDUCER SUMMARY

Transducer Type	Image Shape	Steering Technique	Focusing Technique
B-scan	rectangular	human	conventional
mechanical	sector	mechanical	conventional
linear switched	rectangular	none	conventional
linear phased	sector	electronic	electronic
annular	sector	mechanical	electronic
convex switched	blunted sector	none	conventional
convex phased	blunted sector	electronic	electronic
vector	trapezoidal	electronic	electronic

CRYSTAL MALFUNCTION

B-scan - If a single crystal malfunctions, the entire image is lost.

Mechanical transducers - If a single crystal malfunctions, the entire image is lost.

Linear Switched and Convex Switched Arrays - Malfunction of a single crystal results in drop-out of a single scan line originating at the crystal and extending deeper.

Linear Phased, Convex Phased Arrays, and Vector Arrays - Malfunction of a single crystal results in erratic steering and focusing. The extent to which this affects image quality is variable.

Annular Phased Arrays - Malfunction of a single crystal results in drop-out of a horizontal section of the image at a particular depth. Each crystal has its own unique focal depth. It contributes information only for that depth.

REVIEW - DISPLAYS AND IMAGING

1. The upward deflection of a dot on a screen is characteristic of _____ mode ultrasound displays.

2. The only "mode" display that relates to time as well as position is _____ mode.

3. The "mode" that images anatomy at a particular depth in the body is _____ mode.

4. All of the following are true of B-scanning except:

    a) difficult to perform   b) insensitive to motion
    c) good image quality   d) can make large images

5. True or False? Mechanical scanning produces pictures that are shaped similarly to phased array images.

6. True or False? There are many active elements firing almost simultaneously in a mechanical scanner.

7. True or False? There are many active elements firing almost simultaneously in a phased array scanner.

8. The firing pattern that steers a beam from a phased array transducer up or down relates to _____ .

9. The firing pattern that focuses an ultrasound beam from a phased array transducer relates to _____ .

10. True or False? There are large time delays in the firing pattern of a phased array transducer.

11. True or False? A machine that displays both A-mode and two-dimensional imaging is called a duplex scanner.

ANSWERS - DISPLAYS AND IMAGING

1 -   A-mode or amplitude mode

2 -   M-mode or motion mode

3 -   C-mode or constant depth mode

4 -   b) insensitive to motion.

5 -   True.

6 -   False.

7 -   True.

8 -   electronic slope

9 -   electronic curvature

10 - False: There are SMALL time delays in the firing pattern of a phased array transducer.

11 - False: A machine that displays both DOPPLER and two-dimensional imaging is called a duplex scanner.

TEMPORAL RESOLUTION

TEMPORAL RESOLUTION is the ability to accurately determine the
position of an anatomic structure at a particular instant in time.
This is especially relevant in the imaging of rapidly moving
structures.

Temporal resolution depends upon two factors:
- The extent to which the structure moves.
- The number of images that are created each second: frame rate.

Achieving adequate temporal resolution when imaging a relatively stationary
organ, such as the liver, is not a challenge because the frame rate
can be low. However, to accurately determine the position of
cardiac anatomy in an infant with a typical heart rate of 150 beats
per minute poses a substantial challenge related to temporal
resolution. This requires a high frame rate.

TIME is the resource that must be managed:

Sound travels 1,540 m/s or 154,000 cm/s in the body. This means that
pulses can travel *to* and return *from* a depth of 77,000 cm each
second.

Imaging with a single pulse to a specific depth in order to create a single
scan line requires time.

Imaging with multiple pulses to various depths to create a single
composite scan line requires even more time. (Multi focus and
annular arrays can do this.)

Creating a single frame with a large number of scan lines requires time.

Presenting many frames in rapid sequence requires time.

Therefore, *imaging depth, number of pulses (foci) per line, lines per
frame* and *frame rate* duel for time.

A compromise must be met to balance these factors. The degree of
compromise depends upon the clinical scenario.

Importance of each factor:

Imaging depth - A clinically significant organ must be imaged regardless of
whether it lies superficially or deep in the body. Therefore, in
order to visualize the anatomy, the sonographer must adjust the
system's maximum imaging depth. The deeper the system
images, the longer the listening time for each pulse.

Imaging depth is controlled by the sonographer.

Deeper imaging results in:
* longer listening time.
* a longer pulse repetition period.
* a lower PRF.
* more time required for each imaging scan line.

Multiple focal zones - An ultrasound pulse has only a single focal zone. The
focal zone is the region within the beam that provides the finest
lateral resolution. By using multiple sound beams with different
focal depths to create a single image line, the lateral resolution is
optimal at all depths. This results in superior image quality. The
more foci per image line, the more pulses per image line.

Multiple focal zones are controlled by the sonographer and are
used with phased array transducers (linear, curved, and
annular) only.

More foci per image line result in:
* more sound pulses per line.
* superb lateral resolution at all depths.
* more time per image scan line.
* more time needed to create a frame.

Line density - relates to the number of scan lines that create a single image.
Line density is often set automatically by the ultrasound system
and is not controlled by the sonographer. For a sector scan, line
density is reported in lines per degree. For a rectangular scan,
line density is reported in lines per cm. The greater the line
density, the more pulses per image sector.

Increased line density results in:
- greater detail within the image.
- less "space" between image lines.
- more sound pulses per image.
- more time needed to create a frame.

Frame rate is the number of images created per second.

Units: Hz, frames per second.

Frame rate is determined by the US system and is not directly controlled by the sonographer. When a rapidly moving structure is imaged at an unsuitably low frame rate, the images appear "herky jerky" like an old Charlie Chaplin movie.

More frames per second result in:
- greater accuracy in locating moving structures.
- less time allocated to make each frame.
- decreased line density.

The most common clinical setting is as follows:

1. The sonographer adjusts the system's maximum imaging depth based on the clinical requirements of the exam.

2. The sonographer determines the number of foci per scan line. If superior lateral resolution is desired over a considerable range of depths, then many foci are established. This directly determines the number of sound pulses required to make each scan line.

   Note: Step # 2 is performed only when using a phased array transducer. Phased arrays are the only transducers capable of creating multiple focal zones. One sound pulse is required for each focal zone that is selected.

3. The ultrasound system automatically chooses the frame rate and line density. This choice must carefully balance the goals of temporal resolution (frame rate) and image quality (line density).

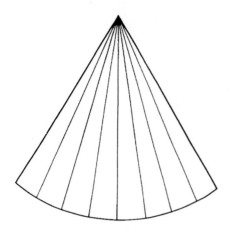 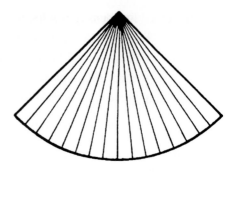

If the frame rate remains constant, there must be fewer scan lines in an image
with a deep field of view (left). The line density will be higher
with a shallow depth of view (right).

With a multi focus system, each scan line is
created from multiple ultrasound
pulses. In this example, five
pulses, each with a different focus,
acquire data from a different
depth. The time required to make
a single frame is long, resulting in
diminished temporal resolution.

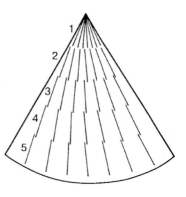

Definitions:

Imaging depth (cm) is the complete depth sound travels per pulse.

Pulses/scan line is the number of trips required for a single image line or
the number of foci per line.

Frames/second is the number of images per second or frame rate.
Typically, frame rate is 10 to 60 images per second.

Pulse Repetition Frequency (PRF) is the number of pulses per second.

Relationships:

$$\textbf{PRF} = \text{pulses/image line} \times \text{image lines/frame} \times \text{frame rate}$$

$$= \text{pulses per picture} \times \text{pictures per second}$$

Pulses/ line **x** imaging depth (cm) **x** frame rate **x** lines/frame $\leq$ 77,000 cm

REVIEW - TEMPORAL RESOLUTION

1. The ability to image accurately is called _____.

2. The ability to accurately determine the position of a structure at any instant in time is called _____.

3. The single most important factor in determining a system's temporal resolution is:
   a) line density
   b) transducer type
   c) imaging depth
   d) frame rate

4. Of the following clinical settings, which is LEAST affected by poor temporal resolution?
   a) Doppler
   b) adult echocardiography
   c) pediatric echocardiography
   d) visualizing a cyst in the liver

5. The factor associated with temporal resolution that a sonographer directly controls while imaging is:
   a) frame rate
   b) line density
   c) lines/frame
   d) imaging depth

6. All of the following are associated with multi focus systems EXCEPT:
   a) phased array transducers
   b) superior lateral resolution
   c) low frame rates
   d) improved temporal resolution

7. What results from # pulses per scan line **x** lines/frame **x** frame rate?
   a) line density
   b) frame rate
   c) pulse duration
   d) PRF

8. True or False? A mechanical scanhead is used to image to a depth of 8 cm. The sonographer increases the imaging depth to 12 cm. The frame rate must be reduced.

9. True or False? A mechanical scanhead is used to image to a depth of 8 cm. The sonographer increases the imaging depth to 12 cm. The line density must be reduced.

10. True or False? A mechanical scanhead is used to image to a depth of 8 cm. The sonographer increases the imaging depth to 12 cm. The line density multiplied by the frame rate must be reduced.

ANSWERS - TEMPORAL RESOLUTION

1 - resolution

2 - temporal resolution

3 - d) frame rate: Temporal resolution is directly related to frame rate. Simply stated, the more frames per second, the better the temporal resolution.

4 - d) visualizing a cyst in the liver: It is reasonable to assume that a liver cyst does not move rapidly.

5 - d) imaging depth: All sonographers adjust the maximum imaging depth of the system. This is the fundamental determinant of frame rate and temporal resolution.

6 - d) improved temporal resolution: Multi focus systems suffer from reduced temporal resolution. Multi focus systems exhibit outstanding lateral resolution by compromising frame rate and temporal resolution.

7 - d) PRF: The number of pulses per line multiplied by the number of lines/frame yields the number of pulses per image. Pulses/image multiplied by frame rate yields the number of pulses per second. This is the pulse repetition frequency.

8 - False: When the sonographer increases the imaging depth, the PRF is reduced. PRF equals pulses per image multiplied by frame rate. The reduction in PRF can be accommodated simply by reducing the number of lines per image (line density) while leaving the frame rate unchanged. In this case, the temporal resolution is maintained at the expense of image quality.

9 - False: When the sonographer increases the imaging depth, the PRF is reduced. PRF equals pulses per image multiplied by frame rate. The reduction in PRF can be accommodated simply by reducing the frame rate while leaving the lines per image unchanged. In this case, the image quality is maintained at the expense of temporal resolution.

10 -True: When the maximum imaging depth is increased, the PRF is reduced. The pulses per frame multiplied by the frame rate is the PRF. Under these conditions, PRF must be reduced.

DOPPLER EFFECT

Doppler ultrasound is used to determine the **speed** and **direction** of blood as it courses through the cardiovascular system.

DOPPLER EFFECT or **DOPPLER SHIFT** is a change in the frequency of sound that results from motion between the source of the sound wave and its receiver. The Doppler shift is the *difference between the frequencies* that are transmitted and received by the transducer after striking red blood cells.

Positive shift:
- Received frequency exceeds transmitted frequency.
- Red blood cells (RBCs) are approaching the transducer.

Negative shift:
- Received frequency is less than the transmitted frequency.
- Red blood cells are moving away from the transducer.

Units: Hertz, cycles per second, units of frequency

Ultrasound systems with Doppler capability are designed by using computers that automatically determine the Doppler shift by comparing the transmitted and received frequencies.

When red blood cells are not moving, the frequency of sound produced by the transducer is equal to the frequency of sound in the reflected echo. The Doppler shift is zero.

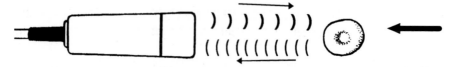

When red blood cells move toward the transducer, the frequency of sound in the reflected wave is *greater* than the frequency initially produced by the transducer.

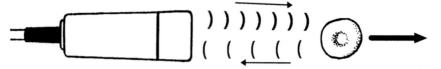

When red blood cells move away from the transducer, the frequency of sound in the reflected wave is *less* than the frequency initially produced by the transducer.

In clinical studies, Doppler shifts range from -10 kHz to +10 kHz.. They are *audible* and result from sound bouncing off red blood cells.

The wave emitted by the transducer is ultrasonic. The wave reflected off the red blood cell is also ultrasonic. Doppler studies generate sound that humans can hear (audible sound) because the *difference* between the two frequencies is within the audible range.

Example: An ultrasound probe emits 5 MHz ultrasound that strikes red blood cells traveling toward the transducer. The reflected wave has a frequency of 5,007,000 Hz. This too is ultrasonic. But, the frequency difference (or Doppler shift) between the waves is 7,000 Hz. A wave with this frequency *can* be heard by man.

As the frequency of a sound wave changes, so too does its wavelength:
- ♦   As frequency increases, wavelength decreases.
- ♦   As frequency decreases, wavelength increases.

> Flow toward transducer ⇨ increased frequency, shorter wavelength.
> Flow away from transducer ⇨ decreased frequency, longer wavelength.

The amount of frequency shift relates to the red blood cell's *velocity*, not speed.

What is the difference between speed and velocity?

Speed is used to describe *how fast* an object is moving with no regard to the direction of travel.

> Velocity has a magnitude and a direction.

Velocity specifies both *the speed and the direction* of travel. Velocity is comparable to an arrow. We must specify its magnitude as well as its direction.

DOPPLER FREQUENCY AND ANGLE

The most accurate velocity measurement is obtained when red blood cells are traveling in a direction parallel to the ultrasound beam. That is, when blood cells travel **directly toward** or **directly away from** the transducer, the Doppler shift measures the actual velocity of the red blood cells.

If, however, there is an angle, Θ, between the direction of blood flow and the direction of the ultrasound pulse, something less than the true velocity is measured.

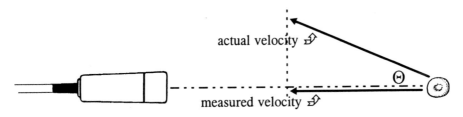

Doppler shift depends upon the **cosine of the angle** Θ between the sound beam and the direction of motion.

measured velocity = actual velocity x cosine Θ

What is a "cosine"?

> Every angle has an associated "cosine." See page 242 for a chart
> of angles and their respective cosines.

With an angle of 0° or 180° between the direction of motion and the sound
> beam, the measured velocity is equal to the true velocity because
> the cosine of these angles is 1.

The cosine of 45° is 0.70. Therefore, if there is a 45° angle between direction
> of flow and the ultrasound beam, only 70% of the velocity is
> measured by the Doppler shift. The actual flow velocity is thus
> underestimated by 30%.

The cosine of 90° is zero. Thus, when the sound beam strikes red blood cells
> moving at a 90° angle, no Doppler shift is measured. For gray
> scale imaging, the optimal angle between the sound beam and
> reflectors is 90°. Thus, Doppler and imaging have distinct and
> conflicting requirements for acquiring optimal data.

Note: When the clinical sonographer merely wishes to identify the presence of
> flow, this interdependence of flow and the cosine is unimportant
> (as long as the incident angle is something other than 90°).
> However, when the sonographer wishes to accurately measure the
> speed and direction of blood flow, the angle between flow and the
> beam should be as close to zero as possible.

THE DOPPLER EQUATION

The **DOPPLER EQUATION** is the mathematical relationship between the frequency shift and the velocity of red blood cells that produce it.

$$\text{Doppler shift} = \frac{2 \times \text{reflector speed} \times \text{incident frequency} \times \text{Cos } \Theta}{\text{propagation speed } + \text{ reflector speed}}$$

The following elements from the Doppler equation are *known*:

incident frequency - This is the frequency of the sound that is emitted by the transducer. **Units**: Hz

propagation speed - This is the speed that sound travels in the body, assumed to be 1,540 m/s.

The **cosine** Θ found in the Doppler equation is *estimated*. On most ultrasound systems, the sonographer can position an "angle correction cursor" on the image that allows the system to estimate the angle between flow and the sound beam. An inaccurate estimate will produce errors in velocity measurements.

The **Doppler frequency** found in the Doppler equation is *measured* by the ultrasound system. The system has electronic circuitry that automatically measures the difference between the emitted and received frequencies. This is the **Doppler shift** (measured in Hz).

When the system has estimated or measured all of these variables, the velocity of the red blood cells (in meters per seconds) is then *calculated* and reported on the screen of the system. Clinicians should be aware that the estimation of the angle Θ is subject to error that may render the velocity measurement inaccurate.

CONTINUOUS WAVE DOPPLER

A continuous wave (CW) Doppler transducer contains two crystals: one receiver and one transmitter. While the transmitter is constantly sending out US energy (duty factor = 1.0), the receiver is continuously listening.

Continuous wave sound cannot be used for imaging. Only pulsed sound can create images. Thus, one of the weaknesses of a pure CW system is a lack of anatomic information. Manufacturers have overcome this limitation by combining two-dimensional imaging (using pulsed wave) with CW Doppler. This optimizes an US system's clinical utility.

Continuous wave's major advantage: CW ultrasound can measure very high velocities.

Continuous wave's major disadvantage: Doppler shifts arise all along the length of the beam. The spectrum produced by a continuous wave system displays Doppler shifts created by any blood cells intersecting the path of the sound beam. This is called **range ambiguity**.

Uni-directional Doppler - a basic system that simply measures the presence of
a Doppler shift. It cannot distinguish whether blood is flowing
toward or away from the transducer. This form of processing is
sometimes called "non-coherent." Often, the Doppler signal is
fed into a speaker that allows the sonographer to listen to the flow.

Bi-directional Doppler - a more sophisticated ultrasound system that
distinguishes between positive and negative Doppler shifts. The
positive and negative Doppler signals are processed independently.
Flow toward and away from the transducer are identified. If the
Doppler signals are presented in an audio format, stereo
headphones or speakers are used. One ear hears flow approaching
the transducer; the other hears flow retreating. "Phase
quadrature" processing is used for bi-directional Doppler.

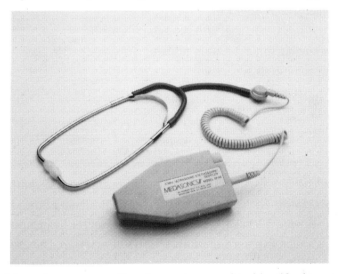

A hand-held continuous wave Doppler system used to identify the presence of
flow. Both earpieces of the stethoscope are attached to a single, small speaker,
indicating that this is a uni-directional Doppler system. (Source: Medasonics)

PULSED WAVE DOPPLER

In **PULSED WAVE DOPPLER,** a **single crystal** in the transducer alternates between sending and receiving ultrasound pulses.

The sonographer positions a cursor within the two-dimensional image. The position of the cursor identifies the region where Doppler shifts and blood velocities are to be measured. This location is called the **sample volume** or **gate**.

The US system calculates the time required for a pulse to travel to the gate and back to the transducer. It then emits an ultrasound pulse and becomes deaf to returning echoes. At the exact moment when echoes would return from the gate, the system listens for a reflection. Only the signal obtained at this instant, corresponding to the location of the sample volume, is processed.

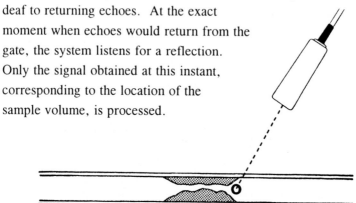

Echoes are analyzed only from the area being investigated, the **sample volume** or **gate**. The sonographer chooses the location of the sample volume. The ability to identify a location for Doppler processing is called **range resolution**.

The time-of-flight (go-return) calculation is as follows:

go-return time = depth of sample volume x 2 / propagation speed

for soft tissue: go-return time (μs) = depth (mm) / 0.77

Since ultrasound systems use pulsed waves to create images, the combination of pulsed Doppler technology and imaging is easily achieved. The ultrasound system simply alternates between imaging and Doppler to help guide the sonographer's sample volume placement.

Note: The best images are obtained with an incident angle of 90°, while
       Doppler is optimized at 0° (parallel to flow). Thus, it is unlikely
       to have simultaneous optimization of flow and image data.

Machines displaying images and Doppler simultaneously are called **DUPLEX**.

Pulsed Doppler's principal advantage is knowledge of the location of the
       blood cells that created the Doppler shift. This is called **range
       resolution**.

Pulsed Doppler's significant disadvantage is the inability to correctly
       measure high velocities. High velocities appear negative. This is
       called **aliasing;** it is the most commonly observed artifact in
       pulsed Doppler imaging.

Aliasing appears only with pulsed Doppler and in the presence of high
       velocities. When Doppler shifts exceeds a value, the **Nyquist
       limit**, velocities are perceived as going in the opposite direction.

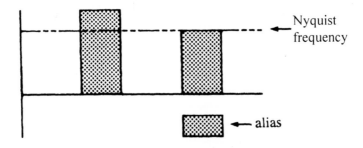

It is possible that you have observed aliasing when attending the movies. The
       wheels of a wagon or an airplane propeller sometimes appear to be
       moving backwards. This is an example of a high velocity in one
       direction appearing as a negative or opposite velocity. A movie is
       subject to the aliasing artifact because it is a pulsed event. Images
       from the movie projector are pulsed onto the screen in a rapid
       sequence in order to give the appearance of motion.

Nyquist frequency (or limit) is the Doppler frequency at which aliasing
       occurs. It is equal to one-half of the pulse repetition frequency.

$$\text{Nyquist limit (Hz)} = \text{PRF (Hz)}/2$$

Note: The depth of the sample volume determines the go-return time required for the pulse to travel away from and back to the transducer. This establishes the number of pulses per second, pulse repetition frequency (PRF). The shallower the sample volume, the greater the PRF, the higher the Nyquist limit, and the less likely aliasing will occur.

Note: The deeper the sample volume is positioned, the lower the Nyquist frequency (the more likely that aliasing will occur.)

A sonographer is less likely to experience aliasing with a low-frequency transducer. Aliasing is more likely to occur with high-frequency transducers. This is another significant difference between imaging and Doppler. When imaging, we desire the highest frequency transducer able to penetrate, creating an image with optimal longitudinal resolution. With Doppler, sonographers often choose a lower frequency transducer to avoid aliasing.

To eliminate aliasing, the sonographer should attempt the following:
- Select a transducer with a lower frequency.
- Select a new imaging view that retains the sample volume within the clinical region of interest, but at a shallower depth.
- Use continuous wave Doppler.

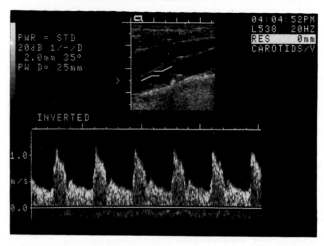

Two-dimensional image and Doppler spectrum from a carotid artery. This is an example of a duplex scan. Note the position of the sample volume and the presence of the angle correction cursor. (Source: Acuson).

DIFFERENCES BETWEEN PULSED AND CONTINUOUS WAVE DOPPLER

PULSED	CONTINUOUS WAVE
one crystal	two crystals
range resolution	range ambiguity
limit on maximum velocity	unlimited maximum velocity
(Nyquist) (aliasing)	

DIFFERENCES BETWEEN DOPPLER AND IMAGING

DOPPLER	IMAGING
best angle = 0^0	best angle = 90^0
pulsed or CW	pulsed only
lower frequencies	higher frequencies
(avoids aliasing)	(improved longitudinal resolution)

DOPPLER SIGNAL PROCESSING

An echo returning after striking a mass of moving blood cells is a complex signal with many varied Doppler shifted frequencies. This is because each blood cell creates a small reflection. These reflections combine into a complicated signal. Ultrasound systems interpret these complex signals using **spectral analysis**.

Within the physiologic range, variations in red blood cell concentration will not affect the ability to successfully perform a clinical Doppler examination.

Reflections from blood cells are of much lower amplitude than those from anatomic structures. Thus, Doppler transducers must be more sensitive to low amplitude echoes than are imaging transducers.

WHAT IS SPECTRAL ANALYSIS?

Spectrum - a collection, assortment, or array.

Analysis - a study, assessment, or examination.

SPECTRAL ANALYSIS entails breaking down a complex signal into its building blocks. Identifying these components may provide insight into the characteristics of the blood flow that created it.

Early methods of spectral analysis:
- ♦ Zero-crossing detection.
- ♦ Time interval histograms.
- ♦ Chirp z-transforms.

These early methods are now considered simple and are no longer used. All of the information contained in the Doppler signal was not analyzed or reported when using these techniques.

Today, spectral analysis of pulsed and continuous wave Doppler signals is accomplished by **Fast Fourier Transform** (FFT). The speed and accuracy of FFT has rendered the early methods obsolete. FFT is very accurate and reports much more of the information found in the original Doppler signal. (See page 141.)

WHY IS IT NECESSARY TO PERFORM SPECTRAL ANALYSIS?

It is rare for all of the red blood cells within the sample volume to travel at an identical velocity.

Usually, the signal received by the transducer is a combination of many varied Doppler-shifted frequencies. The Doppler signal is an aggregate of signals from numerous blood cells with different velocities.

Spectral analysis helps to identify the individual components of the complex signal.

WHAT IS OBTAINED FROM SPECTRAL ANALYSIS?

The end product of spectral analysis may be a plot or graph of Doppler-shifted frequencies as they appear in time.

Another output may be a color "overlay" that is superimposed on a black and
white two-dimensional image (See page 146 for a discussion of
color flow Doppler.)

WHO PROVIDED THE GROUNDWORK FOR OUR CURRENT METHOD OF SPECTRAL ANALYSIS?

In 1822, Jean Fourier proved that any mathematical function can be
represented by a series of sine and cosine curves.

The output of a spectral analysis contains the same information as the original
signal; it is organized differently, however.

In the mid 1960's, the Fast Fourier Transform was developed to transform an
electrical signal into its Fourier components.

The FFT method is particularly effective in performing spectral analysis on
digital devices such as microprocessors and computers. The
application of FFT to clinical Doppler is straightforward because
modern-day ultrasound systems are essentially computers.

WHAT ARE THE LIMITATIONS OF FAST FOURIER ANALYSIS?

At least two samples per cycle are required to accurately identify a sine or
cosine wave.

The sampling rate used with Doppler sonography is the pulse repetition
frequency (PRF).

Therefore, the highest frequency sine wave that can be determined without
ambiguity is equal to one-half the PRF.

This is called the Nyquist frequency.

This limit on measured frequency relates directly to the maximum velocity of
blood cells that can be identified without ambiguity.

The Nyquist limit (and aliasing) occurs only with pulsed ultrasound.

THE IMPORTANT FEATURES OF A SPECTRUM:

While performing a diagnostic study, the clinical sonographer interprets the spectrum with regard to these features.

In addition, the diagnostic medical sonographer must identify other characteristics of the spectrum that provide insight into the flow of blood in the body.

Timing - In what phases of the cardiac cycle does the spectrum appear?

Duration - How long does the spectrum appear?

Direction - Is blood flowing toward or away from the transducer?

Amplitude - What are the maximum and mean velocities of the spectrum?

Broadening - What range of frequencies is present?

ZERO CROSSING DETECTORS

ZERO CROSSING DETECTION was an early, simplified technique for Doppler frequency analysis. It is not used in modern duplex ultrasound systems.

High-frequency waves oscillate many times each second.

Low-frequency waves oscillate fewer times each second.

Each time a wave cycles, the value of its acoustic parameter tends to cross the baseline (zero line).

Early Doppler signal processors counted, or detected, the number of zero crossings within a time span, thereby estimating the frequencies present in a reflected signal.

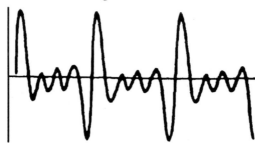

High-frequency waves cross the baseline more often than low-frequency waves

TIME INTERVAL HISTOGRAMS

TIME INTERVAL HISTOGRAMS was another early, simplified (but now obsolete) technique for Doppler frequency analysis. This technique uses zero crossing information to create a plot or graph of the intervals between zero crossings.

The interval between zero crossing relates to the period of the wave. The period relates to the frequency. In this way, the reflected frequencies present in the signal are graphically represented.

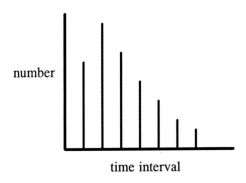

number

time interval

FAST FOURIER TRANSFORM

The current method for pulsed and continuous wave Doppler spectral analysis is by **FAST FOURIER TRANSFORM** (FFT).

FFT uncovers the component frequencies that make up the complex signal returning from the red blood cells. It reveals the individual velocities that are present.

FFT is clinically useful because it is best performed on microprocessors and computers.

Low red blood cell concentrations, anemia, will not limit the ability to perform a Doppler exam.

REVIEW - DOPPLER

1. The Doppler effect is presented as a _____ when the source and the receiver are _____.

2. Starting from the same point, a sound source is moving toward the west at 12 miles/hour and the receiver is moving toward the west at 10 miles/hour. The Doppler shift is _____(+ or -).

3. Starting from the same point, a receiver is moving toward the west at 12 miles/hour and the source is moving toward the west at 10 miles/hour. The Doppler shift is _____(+ or -).

4. Doppler shift produces information about _____.

5. At what angle between the sound beam and the direction of motion will the Doppler shift be a maximum?

6. At what angle between the sound beam and the direction of motion will the Doppler shift be absent?

7. What is the difference between speed and velocity?

8. What is the current method for processing Doppler signals?

9. What is the typical range of Doppler shifts measured in diagnostic imaging examinations?

10. The phenomenon where high velocities appear negative is called _____.

11. The frequency at which aliasing occurs is called _____.

12. The area of interrogation in a pulsed Doppler exam is called _____.

13. The PRF in a 4 MHz pulsed Doppler exam is 5,000 Hz. What is the Nyquist limit?

14. True or False? Pulses made with higher frequency sound are more likely to alias.

15. True or False? The shallower the sample volume is, the more likely a signal is to alias.

16. True or False? Only pulsed wave Doppler exams have a sample volume.

ANSWERS - DOPPLER

1 - frequency shift, moving closer together or further apart

2 - negative: The source and receiver are moving farther apart.

3 - negative: The source and receiver are moving farther apart.

4 - velocity

5 - 0 or 180 degrees: The sound beam and the direction of motion should be parallel.

6 - 90 degrees: There is no Doppler shift because the cosine $90° = 0$.

7 - Speed has only a magnitude. Velocity has magnitude and direction.

8 - The Fast Fourier Transform method of spectral analysis.

9 - between -10 kHz and +10 kHz .

10 - aliasing

11 - Nyquist limit or Nyquist frequency

12 - sample volume or gate

13 - Nyquist limit = PRF/2 = 5,000/2 = 2,500 Hz.

14 - True: Aliasing is more common with higher emitted frequencies.

15 - False: With shallow sample volumes, there is less likelihood of aliasing.

16 - True.

DOPPLER QUESTION

A pulsed Doppler is performed as depicted below. There are five red blood
   cells, labeled A through E, all traveling at a speed of 2 m/s. Their
   direction of travel is indicated by the arrows. Answer the
   following questions.

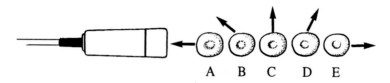

A   B   C   D   E

1. Which two blood cells will produce reflections with frequencies greater
   than the frequency of sound produced by the transducer?

2. Which two blood cells will produce reflections with frequencies less than
   the frequency of sound produced by the transducer?

3. Which RBC will produce the biggest negative Doppler shift?

4. Which RBC will produce the maximum positive Doppler shift?

5. Which blood cell will produce a reflection with the same frequency as the
   sound produced by the transducer (no Doppler shift)?

DOPPLER ANSWERS

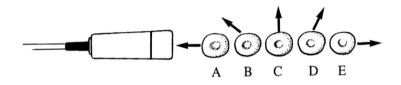

A   B   C   D   E

1 - A and B: Both of these red blood cells are moving closer to the transducer. Thus, both will reflect sound waves with higher frequencies than those of the incident sound. This is a positive Doppler shift.

2 - D and E: Both of these red blood cells are moving away from the transducer. Thus, both will reflect sound waves with lower frequencies than those of the incident sound. This is a negative Doppler shift.

3 - E: This red blood cell is moving directly away from the transducer and will reflect a sound beam with the lowest frequency.

4 - A: In this example, the blood cells moving directly toward the transducer will create the most positive Doppler shift.

5 - C: No Doppler shift is created by red blood cells traveling at a 90^{0} angle to the direction that the sound beam travels. Thus, the frequency of the reflected wave is the same as that for the transmitted wave.

COLOR FLOW DOPPLER

COLOR FLOW DOPPLER is essentially a two-dimensional technique for Doppler. Instead of simply measuring velocities in a single location (the gate of a pulsed Doppler exam) or along a straight line (with continuous wave Doppler), we use color flow Doppler to convert Doppler shifts into different colors. The colors are superimposed on a two-dimensional ultrasound gray scale image. (Examples of color flow Doppler image appear on pages 149 to 152.)

Shades of gray ⇨   structure.

Color   ⇨   flow and function.

Color Doppler is based on **pulsed ultrasound** techniques and is bound by the physics of pulsed waves. Therefore, it provides range resolution at the expense of being subject to aliasing.

Blackness, or an absence of color, on a color flow image indicates either:

- ◆ no blood flow in that particular region, or
- ◆ flow that is normal (90° or perpendicular) to the ultrasound beam.
   This flow geometry does not create a Doppler shift (see page 130).

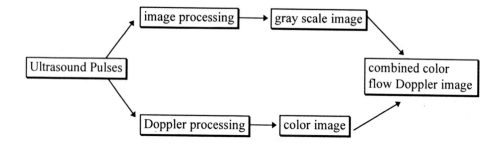

CREATING A COLOR FLOW IMAGE

A conventional pulsed Doppler system acquires Doppler information from a single sample volume, or gate, located on a single scan line.

Color flow systems have many gates positioned at ever increasing depths along many different scan lines.

The frequency shift at each of these multiple locations is measured, thus providing insight into the blood flow velocity at each gate.

The velocity information is encoded into colors, usually red and blue, based on whether the flow is toward or away from the transducer.

COLOR MAPS

Doppler shifts yield information regarding velocity.

Color Flow Doppler (CFD) systems use a "dictionary," or look-up table, to convert velocities into colors. The color is then painted in the location where the Doppler shift was measured. The dictionary is called a **COLOR MAP**, and it appears as a bar or a wheel on the CFD image. (See color flow images on pages 149 to 152.)

Note: Although many laboratories use red to denote flow toward the transducer and blue for flow away from the transducer, there is actually no standardization of color format. Thus, for proper interpretation, one must always study the color map on the image.

The two primary types of color maps are:
- ◆ velocity mode map.
- ◆ variance mode map.

The sonographer selects the type of map used during a clinical exam.

IDENTIFYING AND INTERPRETING COLOR MAPS

To determine the meaning of a color map, follow these steps:

1. Look at the color bar and identify the black region. Blackness indicates that no Doppler shift was measured.

2. Identify the color or colors that are positioned above the black stripe. These colors indicate flow toward the transducer, positive Doppler shifts. If a variety of colors appears above the black stripe, then colors closer to the black stripe indicate lower velocities. Colors further away from the black stripe indicate higher velocities.

3. Identify the color or colors that are positioned below the black stripe. These colors indicate flow away from the transducer. If a variety of colors appears below the black stripe, then colors closer to the black stripe indicate lower velocities. Colors further away from the black stripe indicate higher velocities.

4. Perform this additional step to determine whether the color map is a "velocity " or "variance" mode.
   (Note: At this time, this may have little or no significance to the less experienced sonographer. The importance of different maps is explained in the following pages.)

   At a particular location above or below the black band, determine whether the color of the map changes from side-to-side. Velocity mode maps have only a single color from side-to-side. Variance maps change color from side-to-side.

Refer to color maps on page 150.

COLOR IMAGES AND MAPS

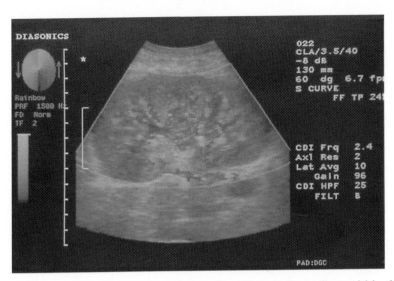

Color flow Doppler image of a kidney. The colors indicate flow within the renal vasculature. Gray scale identifies renal anatomy. (Source: Diasonics, Inc.)

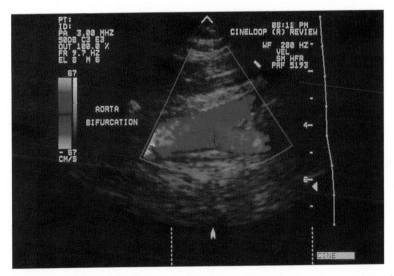

Color Doppler image of the aortic bifurcation. Blood flow is from left to right on the image. The flow color coded in blue indicates flow in the downward direction, away from the transducer. (Source: Advanced Technology Laboratories)

Six color maps used for color flow Doppler imaging. Note that the map on the
left is a variance mode map because its colors change from side-to-side. The
five maps to the right are velocity mode maps. (Source: Toshiba)

Color Doppler map. Refer to the question on page 157. (Source: Toshiba)

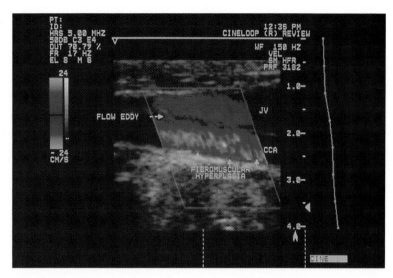

Color flow Doppler image of the carotid artery and jugular vein using a
velocity map. The arterial flow is from right to left on the image. Venous
flow is from left to right. (Source: Advanced Technology Laboratories)

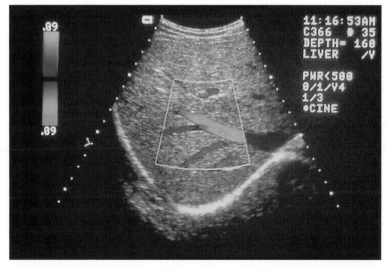

Color Doppler image of hepatic veins using a velocity map. Flow in the small
hepatic veins (red) are flowing towards the transducer and drain into the larger
blue vein, where blood is flowing downward and to the right. (Source: Acuson)

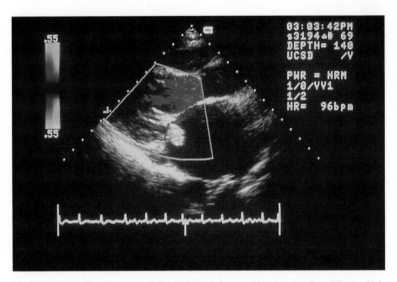

Color flow Doppler image of the heart using variance mode. The solid red
region identifies laminar flow. The smaller light blue and yellow region below
is an area of turbulent flow. (Source: Acuson)

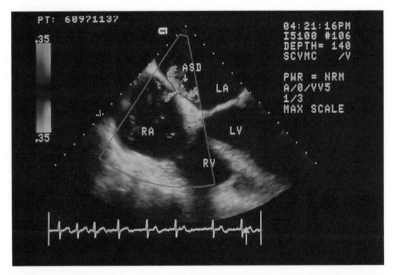

Color Doppler image of the heart using variance mode. A small flow jet
through the atrial septum is seen. The colors in the jet are from the right side of
the color map, identifying the flow as turbulent. (Source: Acuson)

DOPPLER PACKETS

In reality, it is impossible to accurately measure the velocity of red blood cells by using only a single ultrasound Doppler pulse. Actually, information from a number of pulses is required to estimate the velocity of blood cells at a particular gate. This group of ultrasound pulses is called a pulse train or **PACKET**. The pulses in a packet are emitted by the transducer in rapid sequence.

The accuracy of Doppler velocity measurements depends upon the number of pulses in the packet, the **packet size**. The more pulses in the packet, the more accurate is the velocity measurement. However, as packet size increases, more time is required to acquire Doppler information. Color flow systems must therefore balance the desire for accurate Doppler with the need for a diagnostic frame rate.

Typically, packets are composed of between 3 and 20 pulses. Too small a packet (fewer pulses) leads to inaccurate flow measurements. Too large a packet (many pulses) leads to unacceptably low frame rates. Either situation severely limits the clinical utility of a Doppler exam.

When a packet has more pulses, the following result:
- The velocity measurement is more accurate.
- More time is required to collect data from each scan line.
- More pulses are needed to make a single frame.
- The frame rate is lower.
- The temporal resolution is reduced.

VELOCITY MODE

A packet, made of many pulses, radiates down each scan line.

Each pulse produces a single velocity estimate at many gates along the scan line. Many pulses produce multiple velocity estimates for each gate.

With **VELOCITY MODE,** all of the measured velocities for each gate are averaged and the information is consolidated into a single number. The ultrasound system then determines the color corresponding to that average number. Thus, a mean velocity is measured by velocity mode.

See page 148 to learn how to identify a velocity mode map and page 150 for examples of velocity mode color maps. Two examples of velocity mode color Doppler images are found on page 151.

What do the different colors on the image tell us?

We must look at the color bar found in the display. If the color on our image matches one found on the upper half of the color bar, then blood is flowing toward the transducer. The higher the position on the color bar, the greater the velocity of the blood cells.

If the color on our image matches a color on the lower half of the bar, then blood is flowing away from the transducer. The lower the position on the color bar, the greater the speed of the blood cells moving away from the transducer.

An absence of color, or blackness, indicates either no flow or flow that is directed at a 90° angle to the path of the sound beam.

VARIANCE MODE

With **VARIANCE MODE,** the average velocity is calculated (as in the
velocity mode). Then, the variability between the individual
velocity estimates in the packet is examined.

Variance mode has a color map that varies *side-to-side.* See page 148 to learn
how to identify variance mode color maps and page 150 for
examples of variance mode maps. Two examples of variance
mode color Doppler images are found on page 152.

The ultrasound system looks up the color assigned to the average velocity.
Then, if a broad range of velocities is present in the packet,
another color signifying variability is added to the picture. Typical
colors for variability are yellow and green. The wider the spread
or variability of Doppler shifts in the packet, the more the
variance color appears.

Thus, variance mode allows color flow Doppler systems to identify flows that
change from instant to instant. This is turbulent flow.

What does the "green or yellow" mean?

Absence of green: When similar velocities are measured by all of the pulses in
the ultrasound packet, there is little change in the velocity of the
blood cells from instant to instant. The flow is uniform and
smooth and is called *laminar.* Colors within the image are from
the left side of the color map.

Presence of green: From instant to instant, the blood cells are traveling at
different speeds and directions. The flow is chaotic and irregular
and is called *turbulent.* Colors designating turbulent flow appear
on the right side of the color map.

With a variance map:
- left sided colors = laminar flow.
- right sided colors = turbulent flow.

Turbulent flow is **disturbed flow**. It is often present when blood rushes by an irregularly shaped or partially obstructed vessel. The velocity of red blood cells is changing rapidly from instant to instant. Turbulence is often associated with an underlying pathological state. To our ears, turbulence takes the form of a murmur or bruit. To our fingertips, it takes the form of a thrill.

SPECTRAL ANALYSIS OF COLOR FLOW DOPPLER

Color flow Doppler acquires an enormous amount of data related to blood velocities. The huge quantity of data makes it impossible to use the Fast Fourier Transform technique of spectral analysis used for pulsed and continuous wave Doppler. It just isn't fast enough!

Therefore, the technique of choice for spectral analysis of color flow Doppler is **autocorrelation** or **correlation function**.

- Autocorrelation has the advantage of being faster than the FFT technique.
- Autocorrelation has the disadvantage of being less accurate than FFT processing.

QUESTIONS - COLOR MAPS

Answer the following questions based on the color map that appears in the
lower panel of page 150.

1. Which colors indicate flow toward the transducer?

2. Which colors indicate flow away from the transducer?

3. Which colors indicate the absence of a Doppler shift?

4. Is this a velocity mode or variance mode color map? Why?

5. Which colors indicate turbulent blood flow toward the transducer?

6. Which colors indicate turbulent blood flow away from the transducer?

7. Which colors indicate low velocity, laminar flow toward the transducer?

8. Which colors indicate high velocity, laminar flow toward the transducer?

9. Which colors indicate low velocity, laminar flow away from the
    transducer?

10. Which colors indicate high velocity, laminar flow away from the
    transducer?

11. Which colors indicate low velocity, turbulent flow toward the transducer?

12. Which colors indicate high velocity, turbulent flow toward the transducer?

13. Which colors indicate low velocity, turbulent flow away from the
    transducer?

14. Which colors indicate high velocity, turbulent flow away from the
    transducer?

ANSWERS - COLOR MAPS

1 -   All shades of red or yellow appearing in the top half of the color map.

2 -   All shades of blue or green appearing in the bottom half of the color map.

3 -   Black.

4 -   Variance mode: The colors on the right side of the map are different
        from those on the left side.

5 -   The shades of yellow appearing in the top half of the color map.

6 -   The shades of green appearing in the top half of the color map.

7 -   Darker shades of red.

8 -   Lighter shades of red.

9 -   Darker shades of blue.

10 - Lighter shades of blue.

11 - Darker shades of yellow.

12 - Lighter shades of yellow.

13 - Darker shades of green.

14 - Lighter shades of green.

LIMITATIONS OF COLOR FLOW MAPPING

The three major limitations of color Doppler imaging are:

REDUCED FRAME RATES - To create a single two-dimensional color flow image, the system sends out far more ultrasound pulses than are required for a two-dimensional image. This limits the number of frames produced per second (**reduced temporal resolution**).

The four characteristics that reduce frame rates are:
- packet size.
- depth of the region where color flow Doppler data are obtained.
- width of the color region.
- line density, the number of lines in the color region.

ALIASING - High velocity jets occasionally appear as flow in the opposite direction. (Just as a wagon wheel sometimes appears to move backwards on the movie screen.) Color Doppler is subject to aliasing, wherein high velocities are incorrectly displayed as negative.

TOMOGRAPHIC - Ultrasound images a single, thin slice of the target organ. Be careful when considering a single slice to be representative of the entire organ. Additionally, color flow provides information on blood flow only within the imaging plane. There is no direct knowledge of flow characteristics above or below the plane.

ROLES FOR DOPPLER MODALITIES

Continuous wave Doppler identifies high-velocity jets anywhere along the length of the ultrasound beam. Aliasing artifact is impossible.

Pulsed wave Doppler accurately identifies the location of flow (range resolution). It has high frame rates and good temporal resolution.

Color flow Doppler provides two-dimensional flow information directly on the anatomic image. With experienced sonographers, it is a valuable tool that may increase efficiency and diagnostic accuracy. It is subject to low frame rates and reduced temporal resolution.

QUESTIONS - COLOR FLOW DOPPLER

1. All of the following statements regarding color Doppler are true EXCEPT:
    a) Essentially, it is two-dimensional Doppler.
    b) Blood velocities are encoded into colors.
    c) Color flow is a continuous wave technique that can accurately measure all velocities.
    d) Like pulsed Doppler, color flow measures velocities at particular locations and is subject to aliasing artifact.
    e) Different color maps may be selected by the sonographer to analyze the Doppler data in a variety of ways.

2. True or False? Red on a color flow image denotes arterial blood flow, whereas blue identifies venous blood flow.

3. True or False? Red on a color flow image denotes flow toward the transducer, whereas blue always identifies flow away from it.

4. True or False? A single acoustic pulse can create a single scan line on a two-dimensional image.

5. True or False? One acoustic pulse can acquire sufficient Doppler data along a single scan line on a color flow image.

6. Which of the following is not a disadvantage of a large packet?
    a) accurate Doppler information
    b) low frame rates
    c) reduced temporal resolution
    d) increased time to create a single image

7. What is the most significant problem associated with using a packet composed of two pulses?
    a) The spatial resolution is poor.
    b) The temporal resolution is poor.
    c) The frame rate is clinically unacceptable.
    d) The measured flow rates are inexact.
    e) Colors cannot be created with such a small packet.

8. Which of the following may prevent the detection of turbulent flow?
    a) hearing a murmur
    b) visualizing certain colors on variance map
    c) feeling a thrill
    d) visualizing certain colors on a velocity map
    e) hearing a bruit

9. True or False? It is common to encounter turbulence in a patient with some degree of cardiovascular pathology.

10. True or False? It is common to encounter turbulence in a patient with normal cardiovascular status.

11. Which method of Doppler -- pulsed, continuous wave, or color -- is best suited to accurately measure very high velocities?

12. Which method of Doppler -- pulsed, continuous wave, or color -- is best suited to evaluate a tiny jet of blood that appears for a very short time?

13. Which method of Doppler -- pulsed, continuous wave, or color -- is best suited to evaluate the extent of backward flow through a hole in the valves of the heart?

14. Which method of Doppler -- pulsed, continuous wave, or color -- is best suited to avoid aliasing?

ANSWERS - COLOR FLOW DOPPLER

1 - c: Color flow is a derivative of pulsed wave Doppler. The maximum velocity that can be accurately measured by color flow Doppler is controlled by the Nyquist limit. Velocities higher than this limit will result in aliasing artifact.

2 - False: The colors have no relation to arterial or venous blood.

3 - False: Although this statement *may* be true, it is not *always* true. The color map is selected by the sonographer. The map is not always in the red-toward and blue-away format.

4 - True: For image creation, a single pulse of sound is sufficient to create a scan line.

5 - False: Multiple sound pulses, called a packet, are required to measure flows.

6 - a: Accurate Doppler information is the greatest *advantage* of a packet made of many acoustic pulses. The other three choices are indeed disadvantages of large packets.

7 - d: The number of pulses in a packet determines the accuracy of the Doppler information. A packet with only two pulses is too short to acquire precise flow data.

8 - d: Color Doppler systems operating in the velocity mode have a diminished capacity to present information regarding turbulent blood flow. A variance map is a more appropriate selection to visualize turbulence.

9 - True: Turbulence is often associated with cardiovascular pathology.

10 - False: Turbulence is not routinely observed in normal, healthy individuals.

11 - Continuous wave.

12 - Pulsed wave.

13 - Color flow.

14 - Continuous wave.

COLOR DOPPLER QUESTIONS

A color Doppler is performed as depicted below. There are five red blood cells, labeled A through E, all traveling at a speed of 2 m/s. Their direction of travel is indicated by the arrows. Using the color map on the right, answer the following questions.

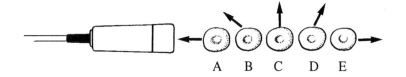

1) Which RBCs will be appear red on the image?

2) Which RBCs will appear blue on the image?

3) Which RBCs will appear black?

Answers:

1 - A and B: The color map indicates that flow toward the transducer will be red. Both blood cells A and B are flowing toward the transducer.

2 - D and E: The color map indicates that flow away from the transducer will be blue. Both blood cells D and E are flowing away from the transducer.

3 - C: The angle between the direction of flow and direction of the sound beam is 90°. The cosine of 90° is 0. Thus, no Doppler shift will be created, and no color will appear on the image.

INSTRUMENTATION

An **ULTRASOUND SYSTEM** comprised all the components necessary to
produce sound beams, retrieve the echoes, and produce visual and
audio signals.

Ultrasound systems process the following information:
- Time of flight.
- Strength.
- Direction (for compound or B-scanning).
- Frequency (for Doppler).

The six electrical components of the system are connected together so that
information can be transferred to and from each individual part.

A **master synchronizer** coordinates all of the components of an ultrasound
system. Some sonographers do not consider the synchronizer to
be a separate component. Whether or not specifically identified,
all of the modules of the system must function in a coordinated
manner.

A **transducer** converts electrical energy into acoustic energy during
transmission, and turns returning acoustic energy into electrical
energy during reception.

A **pulser** (transmitter) controls the electrical signals sent to the transducer
for sound wave generation. The pulser determines the following:
PRF, pulse amplitude, pulse repetition period, the firing pattern
for phased array systems, and the frequency for CW systems.

A **receiver and image processor** receives the electrical signal produced by
the PZT from the returning echoes and creates an image for
presentation on an appropriate display.

A **display** is associated with the visual presentation of the processed data.
The display may be a CRT (television), a transparency, a spectral
plot, or a number of other formats or devices.

Storage devices or "media" are used to permanently archive the US studies.
Storage material consists of video tape, paper, film, transparent
film, and computer discs.

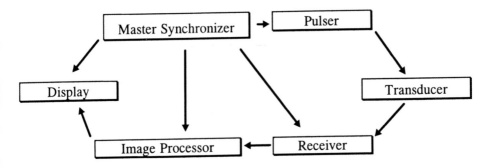

Information flow through an ultrasound system.

PULSER

The **PULSER** receives a timing signal from synchronizer and produces
electrical voltage, ranging from 10 to 500 volts, that travels down
the transducer's wire. The voltage spike stimulates the
piezoelectric crystal.

Greater electrical voltages produce pulses with greater ultrasonic intensities.

The type of system and transducer determine the signal generated by the
pulser. The pulsers for continuous wave, single crystal, and array
transducers are substantially different.

PULSER MODES

Continuous wave:
A constant electrical signal produced by the pulser, in the form of
a sine wave, stimulates the PZT.

electrical frequency = ultrasound frequency

Pulsed wave, single crystal:
A short duration, electrical "spike" smashes into the PZT, which
then vibrates at its resonant frequency.
One electrical spike creates one ultrasound pulse.

Pulse wave, arrays:
Short duration electrical "spikes" stimulate the many crystals
that make up the array.

The number of electrical spikes is the same as the number of transmitting elements in the array.

A **single acoustic pulse** is created from **numerous electrical spikes** -- one electrical spike for each transmitting crystal in the array. The pattern of the spikes serves to focus and steer the acoustic pulse. Voltage ranges from 10 to 500 volts.

TRANSDUCER OUTPUT

The transducer's output is the initial strength of a sound wave. The transducer output controls the amount of ultrasonic energy per pulse that is directed into the patient. Output is controlled by sonographer.

Transducer output is also called: output gain, acoustic power, pulser power, energy output, transmitter output.

Output is dependent upon excitation voltage from the pulser. PZT crystals vibrate with a magnitude related to pulser voltage. The greater the voltage from the pulser, the stronger the acoustic output. The lower the pulser's voltage, the weaker the acoustic output power.

Oftentimes, the output power controlled by the sonographer is measured and reported in decibels. Recall that decibels are a relative measurement (see page 48). In this case, the transducer's actual output is related to the maximum possible output allowed by the design of the ultrasound system.

For example, visualize a system in which output power is measured as 0 dB, -3 dB, or - 6 dB. When the sonographer sets the output power to 0 dB, the system will output 100% of the maximum power established by the manufacturer. When the sonographer adjusts the output power to -3 dB, then the output power is only one-half of the maximum (recall the -3 dB means one-half). When the sonographer adjusts the output power to -6 dB, then the output power is one-fourth maximum. Thus, the sonographer is unaware of the true power output. He is aware of the output power only in relation to the system's maximum allowable output power. Note: This is why amplitude, power, and intensity may be measured in dB (a relative measurement) as well as in temperature, pressure, density, distance, watts, or watts/cm^2 (absolute measurements).

RECEIVER

The **RECEIVER** accepts the small voltages produced by the transducer as it
responds to reflected echoes. Ultimately the signals are processed
and prepared for display on a CRT. The receiver performs five
functions in the following order:

- Amplification
- Compensation
- Compression
- Demodulation
- Reject

> The sequence of receiver operation
> is in alphabetical order.

AMPLIFICATION enlarges the returning signals, so that they may be further
processed by the other electronics in the receiver.

Units: dB. Positive dB since the signal is getting bigger.

Signals from the transducer to the receiver are in the micro- or millivolt range.
They must be boosted to a greater magnitude.
Typically, power is amplified 50 to 100 decibels
(100,000 to 10 billion times larger).
Amplification is also called: receiver gain, overall gain.
Amplification is controlled by the operator.

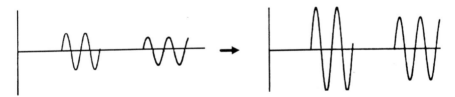

When the sonographer uses too little amplification, some of the processed
echoes will be too small and will not appear on the image.

When too much amplification is applied, all echoes in the image appear
"super-bright" (called **saturation**) and the ability to distinguish
different tissues by visualizing varying echo brightness is lost.

Note: The terms amplification and attenuation are antonyms (opposites).

> Problems resulting from incorrect levels of amplification often affect the entire
> image rather than just a portion of the image.

COMPENSATION - Since attenuation is directly related to the path length, an echo returning from a great depth has a lower amplitude than one returning from a shallow depth. Compensation makes all echoes *arising from similar structures* appear at the same brightness, regardless of the depth of the reflector.

Also called:
- Time gain compensation (TGC).
- Depth compensation (DGC).
- Swept gain compensation.

Objective: Display identical reflectors similarly.

Obstacle: Identical reflectors produce different strength echoes when they are located at different depths.
The go-return distance that the pulse travels is different.
Shorter paths have less attenuation and stronger reflections.
Longer paths have more attenuation and weaker reflections.

Solution: Vary amplification according to depth to adjust for different reflector depths and attenuation. Apply extra amplification for deeper echoes.

TGC is adjusted by the sonographer to optimize image quality.

When TGC is properly adjusted, all images of similar reflectors located at different depths in the body will appear identical.

Note: More TGC is applied with high-frequency sound waves because they experience greater attenuation. Less TGC is applied with low-frequency sound waves since they undergo lesser attenuation.

Problems resulting from incorrect compensation settings often affect only a portion of the image at a particular depth, rather than the entire image.

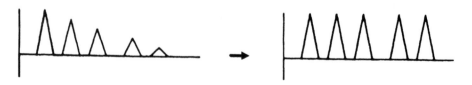

The **TGC CURVE** displays the amount of compensation as related to depth.

Near gain: Minimum amplification level.

Delay: Echoes from shallow depths do not need extra amplification. Regulates the depth at which the compensation begins.

Slope: In this range, extra amplification is applied for ever-increasing depths.

Knee: At this depth and greater, there is maximum and constant amplification of reflected echoes. The system no longer has the capacity to further amplify the signal.

Far gain: Maximum amplification level.

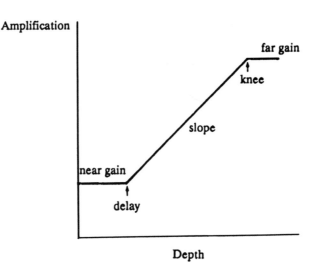

COMPRESSION is the process of reducing the total range of signals, from the smallest to the largest. Compression reduces the difference between the largest and smallest voltages in the signal. Compression is performed without altering the relationships between voltages: the largest signal remains the largest, and the smallest remains the smallest.

Dynamic range is the measure of the *range of signals* that can be processed by the various components of an ultrasound system. **Units**: dB.

Reflected echoes have a large dynamic range, much larger than the
    components of an ultrasound system can manage. Thus, the
    signals cannot be properly processed in their current state and
    must be compressed.

- ◆ The goal of compression is to adapt signals to the confines of the
    electronic components of the ultrasound system.
- ◆ The solution is to reduce the dynamic range, or the extent of voltages
    in the incoming signal.
- ◆ The relative strengths of the signals are maintained:
        The largest signal is still the largest.
        The smallest signal is still the smallest .

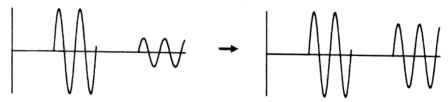

REJECT processes very low-level signals within the reflected echoes that are
    only associated with "noise." These low-level signals detract
    from the clinical utility of the image. Rejection eliminates all
    signals that are below a minimum strength, thereby removing this
    noise. It is controlled by the sonographer.

Reject is also called:
- ◆ Threshold.
- ◆ Suppression.

reject ⇨
level

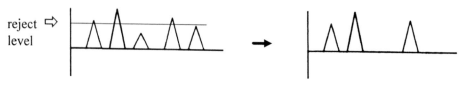

Problems resulting from incorrect threshold levels often affect the entire
image, rather than just a portion of the image.

DEMODULATION changes the shape of the electrical signal into a form that matches the requirements of an input signal to a CRT. It has two steps: rectification and smoothing.

Rectification turns all of the negative voltages into positive ones.

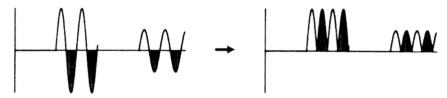

Smoothing places an envelope around the "bumps," to even out the rough edges.

SUMMARY - RECEIVERS

Functions controlled by   ◆ Amplification
the sonographer: ◆ Compensation
       ◆ Reject

Functions automatically   ◆ Compression
set by the system: ◆ Demodulation

Note: The functions of a receiver are performed by all ultrasound systems; however, manufacturers name them differently. After you understand these functions, you can relate them to your own system.

Additionally, some manufacturers borrow the names of functions described above to specify other tasks. This may lead to some confusion when using a particular manufacturer's system in the clinical environment.

QUESTIONS - INSTRUMENTATION

1. All of the following are components of an ultrasound system except:
   - a) transducer
   - b) pulser
   - c) alternator
   - d) synchronizer
   - e) display
   - f) receiver

2. This type of pulser generates a single electrical spike, which ultimately creates a single sound pulse:
   - a) pulsed wave, phased array
   - b) continuous wave
   - c) pulsed wave, single crystal

3. This type of pulser generates a constant electrical signal in the form of a sine wave:
   - a) pulsed wave, array
   - b) continuous wave
   - c) pulsed wave, single crystal

4. This type of pulser generates numerous electrical spikes, which ultimately create a single sound pulse:
   - a) pulsed wave, phased array
   - b) continuous wave
   - c) pulsed wave, single crystal

5. The acoustic power of a sound beam emitted from a transducer is determined by the _____ of the pulser's signal.
   - a) PRF
   - b) frequency
   - c) voltage
   - d) rectification

6. All of the following are functions of the receiver except:
   - a) demodulation
   - b) amplification
   - c) suppression
   - d) attenuation
   - e) compensation

7. Place these functions in the order in which the US system performs them:
   - a) reject
   - b) demodulation
   - c) amplification
   - d) compression
   - e) compensation

8. T or F? Amplification processes all reflected signals in a similar manner.

9. T or F? Compensation processes all reflected signals in a similar manner.

10. When compensation is properly adjusted:
    - a) all reflections on the image appear similar.
    - b) identical reflectors appear on the image similarly.
    - c) identical reflectors have varying brightness on the image, depending on their depth.

ANSWERS - INSTRUMENTATION

1 -   e) the alternator

2 -   c: The pulser of a single crystal pulsed wave system creates a single electrical signal that excites the transducer to create a single acoustic pulse.

3 -   b: The pulser from a continuous wave ultrasound system creates a continuous electrical signal. The electrical signal's frequency determines the frequency of the continuous wave ultrasound.

4 -   a: With phased array technology, the pulser emits numerous electrical spikes that excite numerous crystals in the array. However, only a single acoustic pulse is created from the multiple electronic spikes.

5 -   c) voltage

6 -   d) attenuation

7 -   c, e, d, b, a: (Hint: the order is alphabetical.)

8 -   True.

9 -   False: Those reflections arising from shallow structures are barely compensated. Those arising from deep structures are substantially compensated.

10 - b: When compensation is properly adjusted, identical reflectors similarly appear on the image regardless of the depth of the reflector.

PROBLEMS - COMPENSATION

1. Visualize this clinical situation. You are using a 4 MHz transducer and have adjusted the time gain compensation (TGC) perfectly. This is represented by the solid TGC line below. Now, you are given a new unlabeled transducer with an unknown frequency. The TGC is adjusted to create the same image as that obtained by using the 4 MHz transducer. This new TGC setting is represented by the dotted line. Is the frequency of the unlabeled transducer greater than or less than 4 MHz? Why?

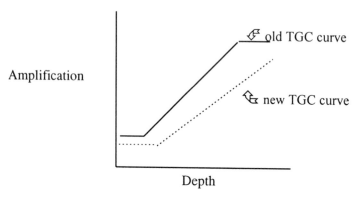

2. Visualize this clinical situation. You are using a 4 MHz transducer and have adjusted the TGC perfectly. The image is optimal. This is represented by the solid TGC line below. You are now given a new, unlabeled transducer with an unknown frequency. The TGC is adjusted to create the same image as that obtained by using the 4 MHz transducer. This new TGC setting is represented by the dotted line. Is the frequency of the unlabeled transducer greater than or less than 4 MHz? Why?

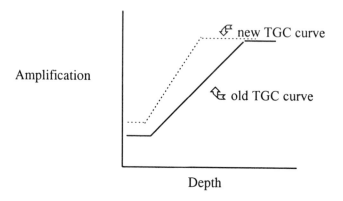

ANSWERS - COMPENSATION

In example #1, the new transducer's frequency is less than 4 MHz. Notice that
the TGC curve is shifted for the new transducer. At any particular
depth in the region of the "slope," less compensation is applied
when using the new transducer. In other words, the sonographer
used a smaller amount of compensation to create the same image.

Recall that compensation is used by the sonographer to pay back or make
up for attenuation. In this example, the decreased compensation
informs us that there was less attenuation. Lesser attenuation
occurs with lower frequency transducers. Thus, the new TGC
curve implies the use of a transducer with a frequency that is less
than 4 MHz.

In example #2, the new transducer's frequency is higher than 4 MHz. Notice
that the TGC curve is shifted for the new transducer. At any
particular depth in the region of the "slope," extra compensation is
applied when using the new transducer. In other words, additional
compensation is applied.

Recall that compensation is used by the sonographer to pay back or make
up for attenuation. In this example, the extra compensation
informs us that there was greater attenuation. More attenuation
occurs with higher frequency transducers. Thus, the new TGC
curve, with its extra attenuation, implies the use of a transducer
with a frequency that exceeds 4 MHz.

QUESTIONS - CLINICAL INSTRUMENTATION

1. Which of the following is the most reasonable initial action to take when the image on your US system displays only reflectors in a region close to the transducer, but displays nothing in deeper regions?
   a) Adjust the system's compensation.
   b) Adjust the system's compression.
   c) Use a higher frequency transducer
   d) Increase the output power.
   e) Adjust the reject level.

2. Which of the following is the most reasonable action to take if the image on your ultrasound system displays only echoes arising from bright reflectors at all depths? However, no weak reflectors appear anywhere on the image.
   a) Adjust the system's compensation.
   b) Increase the system's overall gain.
   c) Use a higher frequency transducer.
   d) Increase the output power.
   e) Adjust the reject level.

3. Which of the following is the most reasonable action to take when the image on your ultrasound system displays only bright reflections, at all locations on the image, that arise from both strong and weak reflectors in the body?
   a) Adjust the system's compression.
   b) Decrease the overall amplification or gain.
   c) Use a higher frequency transducer.
   d) Use a lower frequency transducer.
   e) Adjust the reject level.

4. Which of the following is the most reasonable action to take when the image on your US system displays reflectors only in regions deeper than 5 cm, but displays nothing from shallower depths?
   a) Adjust the system's compensation.
   b) Use a higher frequency transducer.
   c) Decrease the output power.
   d) Adjust the reject level.

ANSWERS - CLINICAL INSTRUMENTATION

1 - a) Adjust the system's compensation: Apply extra compensation to the regions far away from the transducer. The problem is that the system is not displaying echoes arising from reflectors deep in the body. Possible causes for this include: insufficient output power, inadequate amplification, or inadequate compensation. Thus, choices a or d are acceptable. To remedy to this problem, first increase the compensation in order to brighten these deep echoes.

It is improper to increase the output power first because this exposes the patient to extra acoustic power and increases the potential for bioeffects. The general rule is as follows: *Increasing output power should only be attempted after additional amplification or compensation fails to correct a problem with an image.*

2 - e) Adjust the reject level: This system is not displaying weak echoes anywhere. Consistent with this problem is the fact that the reject (threshold or suppression) level of the system is set too high and the system is discarding all lower level echoes. If the sonographer lowers the threshold, more lower level signals will be processed and will be likely to appear in the image.

3 - b) Decrease the overall amplification or gain: In this case, the system is not distinguishing between large and small echoes. Everything appears bright. This is an example of saturation. Saturation is corrected by decreasing the receiver gain. This will enhance the system's ability to display weak echoes differently from strong echoes.

4 - a) Adjust the system's compensation: The system is not displaying echoes arising from superficial structures. However, it is adequately processing deeper echoes. When the sonographer faces a problem that is related to brightness and the depth of the reflector, he should attempt to adjust compensation. (Recall that compensation adjusts for varying path lengths.) Boosting the compensation for the region close to the transducer is likely to correct the defect in this image.

DISPLAYS

A **CATHODE RAY TUBE** (CRT) is a television, or a large vacuum tube.

The inner surface of the CRT screen is coated with **phosphors that glow** when struck by electrons.

An electron gun, located at the back of the tube, fires electrons (tiny charged particles) toward the screen.

The electrons travel through rapidly varying magnetic fields that cause the electron beam to sweep across the CRT screen. The magnetic fields are created by deflecting coils or plates located in the "neck" of the CRT.

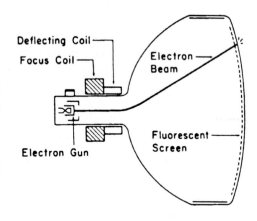

The electrons strike the phosphors on the screen, which stimulates them to emit light. The brightness of the spot on the screen is directly related to the voltage, or strength, of the electron beam.

The image presented on a standard domestic television display is a result of a collection of 525 closely spaced horizontal lines. First, lines 1, 3, 5, 7 ... 525 are written, by sweeping the electron beam

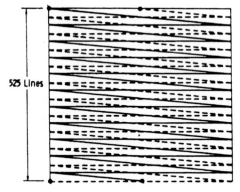

across the screen (solid lines). The electron beam is located at the upper left hand corner and swept horizontally across the screen.

The beam is then displaced downward and swept horizontally across the screen again. This process is repeated 263 times, and the result is an image called the odd-field.

Then the remaining, interspersed lines 2, 4, 6, 8 ... 524 are written (dashed lines). This is called the even field. The combination of an even field and an odd field is called a **frame**. The process of odd- and even-field image presentation is called an **interlaced display**.

On normal televisions, each field is created in 1/60th of a second. Therefore, each frame requires 1/30th of a second, and 30 frames are painted on the CRT screen each second. A TV's frame rate is 30 Hz.

Some displays are not interlaced. They simply paint the image lines in sequence from top to bottom: lines 1, 2, 3, 4, 5 ... 524, 525. This form of display is called **non-interlaced** or **progressive scan**.

Bistable - either on or off, white or black. Bistable displays create images that are purely white or black; they contain no shades of gray.

Gray scale displays create images with various levels of brightness (white, dark gray, light gray, black).

Different shades of gray are assigned to different echo amplitudes:
  - Strong reflections are white.
  - Intermediate strength reflections are gray.
  - Weak echoes are very dark gray.

Gray scale displays enhance the clinician's ability to distinguish boundaries, tissue types, and textures.

Alternatively, a color CRT uses different colors on the image to correspond to different echo strengths. This display format is called **B-color**.

Controls for a CRT display:
Brightness - controls the brilliance of the signals displayed.
Contrast - controls the range of brilliance, from weakest to strongest, that is displayed. When two images of the same anatomy are compared, the one with the "blacker" blacks, "whiter" whites, and fewer shades of gray is called the higher contrast image.

SCAN CONVERTERS

SCAN CONVERTERS make gray scale displays possible by storing the image data and then displaying it on a CRT. The image information can be manipulated, or altered, while stored in the scan converter. This technique is called **processing**.

Example: A black-on-white image may be converted into a white-on-black image.

PREPROCESSING describes any manipulation of image data *before* it is stored in the scan converter.

POSTPROCESSING describes manipulation of data *after* it has been stored in the scan converter memory, but prior to display. This increases the versatility of displays, but data that have not been stored are irretrievable.

> To determine whether processing is "pre" or "post" ask the question: "Have the data already been stored in the scan converter?"

ANALOG and DIGITAL

Analog variables can attain a continuum of values:
- Weight of an individual.
- Length of a piece of string.

Digital variables can attain only discrete values:
- Number of children in a family.
- Number of words in an article.

Example: A digital scale measures weight to the nearest pound, whereas an analog scale is able to measure fractions of pounds.

ANALOG SCAN CONVERTER

ANALOG SCAN CONVERTERS represent the earliest form of converter that made gray scale imaging possible. This type of converter is similar to a CRT except that its phosphor screen has been replaced with a panel that captures and stores electrons.

The panel divides the picture into a matrix (up to 1000 x 1000 picture elements) of electrical storage elements (silicon wafers). The matrix has up to 1 million picture elements, which provides superb spatial resolution. An electron gun in the analog converter shoots electrons that are stored in each location within the panel. The stored charges are then read to retrieve the information and display it on a CRT.

Disadvantages of Analog Scan Converters:

- Image fade — charges on silicon wafer dissipate.
- Image flicker — constant switching between read and write modes.
- Drift — inconsistent pictures from day to day.
- Deterioration — tubes age and the image degrades.

These disadvantages, combined with enormous advances in digital technology, have retired the analog scan converter from clinical service.

DIGITAL SCAN CONVERTERS

DIGITAL SCAN CONVERTERS use a *computer* and *computer memory* to digitize images. Digitizing the image converts it into binary numbers (zeroes and ones) and stores them in memory. The numbers can be processed and then re-translated for presentation on the display of the ultrasound system.

Advantages of Digital Scan Converters:

- Uniformity — consistent gray scale qualities throughout the image.
- Stability — absence of fade or drift.
- Durability — not influenced by age or heavy workload.
- Rapidity — near instantaneous data processing.
- Error free — extreme accuracy.

A **PIXEL** is the smallest element of a digital picture. If we divide the picture into a grid (like a checkerboard), each square is a pixel. Each pixel can have only one color or shade of gray.

The more pixels per inch (**pixel density**), the greater the detail in the image and the better the spatial resolution in the picture.

The upper image has a greater pixel density than the lower.

BIT (**B**inary dig**IT**) is the smallest amount of digital computer memory. Computer memory is also called **RAM or random access memory**. A **byte** is a collection of eight bits of computer memory.

A bit is **bistable**; it has a value of either 0 or 1. A group of "bits" is assigned to each pixel to store the gray shade assigned

> The more bits per pixel, the greater the selection of gray shades that can be represented.

to that pixel. The more bits assigned per pixel, the more extensive the choice of gray shades.

A group of bits represents a binary number. For example, 100010 is a binary number made of six bits. Each bit has a value of either 0 or 1. Compare this to decimal numbers that we use in our everyday life, which are made up of digits ranging from 0 to 9.

The image on the left has eight bits assigned to each pixel, whereas the image on the right has only two bits per pixel. More shades of gray appear on the image with the greater number of bits per pixel.

8 bits per pixel   2 bits per pixel

HOW MANY DIFFERENT GRAY SHADES CAN A COLLECTION OF BITS REPRESENT?

1. Find out how many bits are assigned to each pixel.

2. Multiply the number 2 by itself the same number of times as there are bits. That's the answer.

Example: How many shades can be represented by 4 bits?

We have 4 bits, so take the number 2 and multiply it by itself 4 times!

$$2 \times 2 \times 2 \times 2 = 16$$

Sixteen different gray shades are represented by four bits.

Example: How many shades are represented by one bit?

The answer is 2 (off or on, black or white).

This display is bistable.

Example: What is the largest number of shades represented by 7 bits?

We multiply the number 2 by itself 7 times.

$$2 \times 2 \times 2 \times 2 \times 2 \times 2 \times 2 = 128$$

Seven bits can represent 128 different shades of gray.

CONVERTING ANALOG AND DIGITAL SIGNALS

Digital scan converters increase the clinical utility of our ultrasound systems. Thus, ultrasound systems must have the capacity to transform analog signals (which come from the transducer) into the digital form that is required by digital scan converters. Additionally, digital signals must be converted back into analog form, so that they can be displayed on a television screen.

The process of returning echo information is described by the following five steps:

1. Electrical signals sent by the transducer to the receiver are in analog form. In order to place the information into computer memory, the signals must be converted into digital form by using an **analog to digital converter** (A to D).

2. The image information is then stored in the scan converter's computer memory. Any manipulation or processing of the reflected signals prior to this instant in time is called **preprocessing**.

3. The image information, in the form of digital data, continues to be processed by the ultrasound system's computers. Any processing after storage in the digital scan converter is called **postprocessing**.

4. Digital signals cannot be displayed on conventional televisions because TV video signals are analog. Thus, the digital data is converted to analog form by using a **digital to analog converter** (D to A).

5. The signal, now in analog form, is transmitted to the CRT display for presentation.

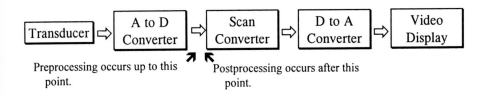

| Transducer | ⇨ | A to D Converter | ⇨ | Scan Converter | ⇨ | D to A Converter | ⇨ | Video Display |

Preprocessing occurs up to this point.   Postprocessing occurs after this point.

EXAMPLES: PRE- AND POSTPROCESSING

Sometimes, it is difficult to determine whether the manipulation of image
information performed by a particular ultrasound system is
preprocessing or postprocessing. If, however, the adjustment is
made on an image that is
freeze-framed, the process
must be postprocessing.

> Manipulation of a frozen image
> must be postprocessing.

Certain forms of processing may be performed either before or after image
storage in the scan converter. Thus, image manipulation that is
preprocessing on one manufacturer's system may be
postprocessing on a another manufacturer's system.

Certain examples of pre- and postprocessing are important to the clinical
ultrasonographer and are worthy of detailed discussion. They are
write magnification, read magnification, and **contrast
enhancement.**

WRITE AND READ MAGNIFICATION

Magnification is a process in which a particular region of an image is selected
by the sonographer and magnified to fill the entire screen.

WRITE MAGNIFICATION, also called write zoom or regional expansion, is
a form of **preprocessing** wherein the zoom technique is applied
during data acquisition and *before* storage in the scan converter.

Four steps are associated with write zoom:

1. The sonographer selects the **region of interest** (ROI) within the image to be
   magnified.

2. The ultrasound system rescans only the region of interest. Thus, new
   information is collected. All of the scan lines generated by the
   transducer are now squeezed into the ROI. The line density is
   increased, and the spatial resolution is improved.

3. The information collected from this limited region undergoes analog to digital conversion and is stored in the scan converter.

4. The information undergoes digital to analog conversion and is displayed on the screen.

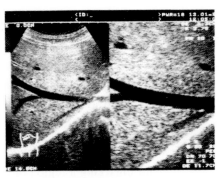

Write magnification. Source: (Toshiba)

The advantages of write magnification:

- The system collects new image data (preprocessing).
- The system scans only the region of interest.
- The system does not scan extraneous regions.
- More scan lines pass through the region of interest.
- Detail is improved in the region of interest.
- More pixels are contained in the ROI, resulting in increased spatial resolution.

READ MAGNIFICATION, (read zoom) is a form of **postprocessing,** wherein the zoom technique is applied *after* data acquisition and *after* the image data are stored in the scan converter.

Five steps are associated with read magnification:

1. The ultrasound system scans the anatomy and creates an image.

2. The information collected from this scan undergoes analog to digital conversion and is stored in the scan converter.

3. The sonographer selects the region of interest to be magnified, but the region is not re-scanned.

4. Pixels in the scan converter associated with the region of interest are enlarged, so that ROI fills the entire screen.

5. The information undergoes digital to analog conversion and is displayed on the video screen.

The advantages of read magnification:

- A zoomed region of interest.
- Larger pixels in the region of interest.

The disadvantages of read magnification:

- The system retains image data from extraneous regions.
- The same number of scan lines appear through the ROI.
- There is a similar amount of detail in the expanded region.
- There is no improvement in spatial resolution.
- The same number of pixels appear in the expanded region.

Thus, the write magnification technique is preferred over read magnification for enlarging a region of interest in a clinical image.

The upper image identifies the region of interest in a non-magnified picture. The image on the lower left was created by using read magnification. Note the large pixels in the image. The image on the lower right was created by using write magnification. Note the small pixel size and greater detail in the image.

Read magnification   Write magnification

CONTRAST ENHANCEMENT

CONTRAST ENHANCEMENT is a **postprocessing** function that translates the digital numbers, stored in the scan converter, into shades of gray or brightnesses on the CRT; it is also called **gray scale mapping**.

Three general forms of contrast enhancement or gray scale mapping exist:
- Linear assignment gray scale mapping.
- High-level gray scale enhancement (or low-level compression).
- Low-level gray scale enhancement (or high-level compression).

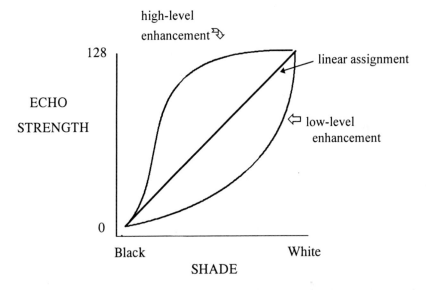

Note: High- and low-level gray scale compression should not be confused with the compression function of the receiver described on page 169. These are entirely different processes.

LINEAR ASSIGNMENT GRAY SCALE MAPPING

Gray scale mapping assigns specific TV display brightnesses to the digital numbers stored in the scan converter. Imagine that the human eye can only distinguish ten different gray shades ranging from black to white. (This assumption, although not perfect, is generally accurate.) Also imagine that the digital scan converter in our ultrasound system has seven bits allocated to each pixel, to store the strength of the reflected signal. Seven bits allows for 128 different echo strengths to be distinguished or identified.

With linear mapping, the 128 echo strengths will be assigned ten gray shades distinguishable by the human eye. Pixels with digital numbers from 1 to 13 will appear black on the screen. Pixels with digital numbers from 14 to 26 will appear dark gray. Pixels with digital numbers from 115 to 128 will appear white. In other words, approximately 13 sequential echo strength numbers in the scan converter will appear as the same shade of gray on the display.

Example of linear mapping:

Echo Strength	Assigned Shade
1 to 13	black
14 to 26	very dark gray
27 to 39	dark gray
40 to 52	dark medium gray
53 to 65	medium gray
66 to 78	light medium gray
79 to 91	dark light gray
92 to 104	medium light gray
105 to 118	very light gray
119 to 128	white

Under some circumstances, linear assignment may present the reflected echo information in a manner that optimizes its clinical utility. However, what if the sonographer needs to see echoes that are close in numerical value as different shades of gray? This occurs when imaging the liver, for example, and attempting to distinguish abnormal masses situated within normal tissues. Linear assignment may not be able to display slightly different tissues (with slightly dissimilar echo values) as different gray shades.

LOW-LEVEL GRAY SCALE ENHANCEMENT

With low-level enhancement, the sonographer adjusts the US system to alter the relationship between the numbers stored in the scan converter and the gray shades displayed on the image. Using our example from above, let us now assign values between 0 and 5 as black, between 6 and 10 as very dark gray, and between 11 and 16 as dark gray. The advantage of this postprocessing scheme is that more shades of gray are assigned to the lower strength echoes. Now, small differences in the lower strength echoes are easier to distinguish visually. Low-level echoes will now appear as different shades of gray when they are only five numbers apart.

But, if six shades of gray are used to display echoes

> Low-level enhancement is useful when trying to distinguish differences in low-level echoes.

ranging from 1 to 30, only four shades remain to display the remaining 98 levels of echo strength. This means that the remaining 98 higher echo levels are compressed into only four shades of gray; hence, the term high-level compression. A broad range of stronger echoes is compressed into fewer gray shades.

Example of low-level enhancement:

Echo Strength	Assigned Shade
1 to 5	black
6 to 10	very dark gray
11 to 15	dark gray
16 to 20	dark medium gray
21 to 25	medium gray
26 to 30	light medium gray
31 to 54	dark light gray
55 to 78	medium light gray
79 to 102	very light gray
103 to 128	white

> For weak echoes, slight differences in echo strength will appear as different gray shades.

HIGH-LEVEL GRAY SCALE ENHANCEMENT

With high-level enhancement, the US system is adjusted to alter

> High-level enhancement is useful for distinguishing differences in high-level echoes.

the relationship between the echo strengths stored as numbers in the scan converter and the gray shades displayed on the CRT. By using our example from above, now assign values between 0 and 25 as black; between 26 and 50 as very dark gray; between 51 and 75 as dark gray; and between 76 and 100 as dark medium gray.

The advantage of high-level enhancement postprocessing is that more shades of gray are assigned to the higher strength echoes. In this example, six remaining shades of gray are assigned to the echoes ranging from 101 to 128. Thus, each range of four echo values will be seen as a different shade of gray. Slight differences in the higher strength echoes will be easier to distinguish visually. This occurs because the high-level echoes will appear as a different shade of gray when they are only four numbers apart. But, since six shades of gray are used for echoes ranging from 101 to 128, there are fewer gray shades to display the remaining 100 levels of echo. Now, the lowest 100 echo levels are compressed into four shades of gray (low-level compression). A broad range of weak echoes is compressed into a few gray shades.

Example of high-level enhancement:

Echo Strength	Assigned Shade	
1 to 25	black	For strong echoes,
26 to 50	very dark gray	slight differences
51 to 75	dark gray	in echo strength
76 to 100	dark medium gray	will appear as
101 to 105	medium gray	different gray
106 to 110	light medium gray	shades.
111 to 115	dark light gray	
116 to 120	medium light gray	
121 to 124	very light gray	
125 to 128	white	

The top image is an example of linear gray scale assignment. The picture on
the lower left has been postprocessed by using a low-level
enhancement technique. Note that slight differences in dark gray,
low-level signals are now visible; for example the flowers in front
of the house. The picture on the lower right has been
postprocessed by using a high-level enhancement technique. Note
that slight differences in bright, high-level areas are now visible;
for example, the area near the garage on the left side of the image.

Low-level enhancement   High-level enhancement

STORAGE MEDIA

Paper media: charts from pen writers
    Advantages: portability
                does not require a device to read
    Disadvantages: bulky, hard to store
                difficult to make copies
                cannot display dynamic images

Magnetic media: computer discs
                computer memory
                magnetic tape
                video tape
    Advantages: able to store large amounts of information efficiently
                can transfer information over phone lines
                can store and play dynamic images
    Disadvantages: many different formats (e.g., VHS and beta VCRs)
                require a machine to view the information

Chemically derived media: photographs
                transparent film
                Matrix camera film
    Advantages: high resolution
                accepted in the medical community
                can produce color images
    Disadvantages: difficult to store and retrieve
                require chemical processing

Optical media: laser discs
                compact discs
    Advantages: store huge amounts of data
                inexpensive
    Disadvantages: require a display system
                no standardized format for image display and storage

It is important to understand the strengths and liabilities associated with each form of data storage. As an exercise, expand the lists provided for each type of system.

QUESTIONS - INSTRUMENTATION

1. All of the following are true of ordinary cathode ray tubes EXCEPT:
   a) They are large vacuum tubes with a phosphor-coated screen.
   b) Electrons are emitted by a gun and swept across the front of the tube by using varying magnetic fields.
   c) There are 600 horizontal scan lines from top to bottom, painted in order from 1 to 600.
   d) There are 60 fields generated each second that combine into 30 frames per second.

2. Which of the following correctly describes a typical CRT?
   a) interlaced   b) progressive scan
   c) bistable   d) non-interlaced

3. Which electronic component is required for gray scale imaging?
   a) VCR   b) non-interlaced display
   c) computer memory   d) scan converter

4. All of the following are disadvantages of analog scan converters EXCEPT:
   a) image fade   b) low-resolution image
   c) image flicker   d) deterioration

5. Which of the following will produce an image with the best quality?
   a) a digital scan converter with 256 x 256 pixels
   b) a digital scan converter with 512 x 512 pixels
   c) a digital scan converter with 128 x 128 pixels
   d) an analog scan converter with 1000 x 1000 array

6. Which of the following statements regarding a bit of computer memory is false?
   a) It is the smallest element of an image.
   b) It is bistable.
   c) Eight bits combine to make a byte.
   d) It is a component of a digital scan converter.

7. How many gray shades can be represented by a group of 8 bits? 4 bits? 2 bits?

8. Which of the following statements regarding a pixel is false?
   a) It is the smallest element of a digitized image.
   b) A collection of bits, assigned to each pixel, stores the shade of gray of the pixel.
   c) It can display up to three gray shades simultaneously.
   d) Image quality is improved when the number of pixels is high.

Typically, are the following forms of information digital or analog?
   9. the signal from the transducer to the receiver
   10. the input signal to the digital to analog converter
   11. the output signal from the analog to digital converter
   12. the input signal from the analog to digital converter
   13. the output signal from the digital to analog converter
   14. a typical video signal

Typically, are the following procedures pre- or postprocessing?
   15. modifying a frozen image
   16. read zoom
   17. write zoom
   18. adjusting the brightness on the CRT
   19. increasing the receiver gain

20. All of the following are characteristics of write magnification EXCEPT:
   a) preprocessing function
   b) more pixels in the region of interest
   c) identical regions stored in the scan converter before and after zoom is enabled
   d) the preferred method of image magnification

21. Sketch a gray scale assignment that can be used to visualize and distinguish high-level echoes that differ from each other only slightly.

22. Sketch a gray scale assignment that can be used to visualize and distinguish low-level echoes that differ from each other only slightly.

ANSWERS - INSTRUMENTATION

1 - c: A normal CRT creates images by using 525 horizontal scan lines that appear on the screen in an interlaced format.

2 - a: A typical CRT uses an interlaced display format.

3 - d: Scan converter technology was used to create gray scale displays.

4 - b: Analog scan converters produced high-resolution images. The other three choices represent shortcomings of analog converters.

5 - d: With respect to spatial resolution, the scan converter with the highest number of picture elements produces the best image. Choice d has 1 million picture elements (1,000 x 1,000) and, therefore, has the highest image quality.

6 - a: A bit is not a part of an image. A bit is the smallest element of computer memory.

7 - To determine the number of gray shades that a group of bits can represent, simply multiply 2 by itself as many times as there are bits.
8 bits can represent 256 different gray shades.
4 bits can represent 16 different gray shades.
2 bits can represent 4 different shades.

8 - c: At any given moment, a pixel can display only a single shade of gray. Think of a pixel as a square on a checkerboard. The coloration is uniform throughout.

9 - analog

10 - digital

11 - digital

12 - analog

13 - analog

14 - analog

15 - postprocessing

16 - postprocessing

17 - preprocessing

18 - postprocessing

19 - preprocessing

20 - c: With write magnification, the system concentrates all of the image
         scan lines into the region of interest (ROI). The anatomy on either
         side of the ROI is no longer scanned.

21 and 22.

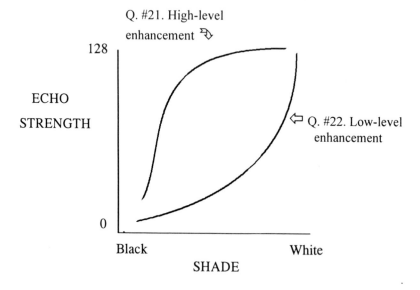

ARTIFACTS

An **ARTIFACT** is *any* error in imaging.

Types of artifacts include reflections that are:
- Not real.
- Not present on the image.
- Incorrect size or shape.
- Improper location or brightness.

Artifacts are caused by:
- Violation of assumptions that are incorporated into the design of ultrasound systems.
- Violation of assumptions of viewer.
- Equipment malfunction.
- Poor ultrasound system design.
- Improper use of ultrasound systems.
- Physics of ultrasound.
- Operator error.
- Viewer error.

Basic assumptions of acoustic imaging system include:
- Sound travels in a straight line.
- Reflections are produced by structures that are located along the main axis of the sound beam.
- Intensity of a reflection corresponds to a reflector's scattering strength.
- Sound travels exactly 1,540 meters/second.
- The imaging plane is extremely thin.
- Sound travels directly to a reflector and back to the transducer.

> Most acoustic artifacts can be explained by identifying the assumptions that were violated during the imaging process.

TYPES OF ARTIFACTS

AXIAL AND LATERAL RESOLUTION

Causes: Ultrasound interacting with tissues.
                  Physics of ultrasound.

Presentation: Closely spaced objects appear as one.
                  Numerous tiny reflectors in the body appear as
                      fewer, larger spots on the image.
                  Too few reflectors appear on the image.
                  The reflectors that do appear are too large.

AXIAL RESOLUTION (see page 90.)

Multiple structures are oriented along the main axis of ultrasound beam.
However, they appear only as one reflector on the image.
Units: mm (any unit of distance). Smaller values produce better pictures.
Synonyms: longitudinal, range, radial, depth.
Approximately equal to half of the length of the ultrasound pulse.
Typical value: 0.3 to 1.0 mm.

> Ultrasound systems that create short pulses have improved image
> quality. The numerical value associated with axial resolution is
> small. This is typically a higher frequency transducer.

LATERAL RESOLUTION (see page 92; and **focusing**, see page 89.)

Multiple structures are perpendicular to the main axis of sound beam.
They appear as one reflector on the image.
Units: mm (any units of distance). Smaller values produce better
    pictures.
Synonyms: azimuthal, transverse, angular.
Approximately equal to beam diameter, varies with depth.
Typical value: 2 to 10 mm (improved by focusing).

> Ultrasound systems that produce narrow pulses have improved image
> quality. The numerical value associated with their lateral
> resolution is small. This is typically a focused transducer.

ACOUSTIC SPECKLE

- Appears on the image as tissue texture close to the transducer, but does not directly correspond to scatterers in tissue.
- Is produced by ultrasound wavelet interference unrelated to anatomy.
- Results in general image degradation.

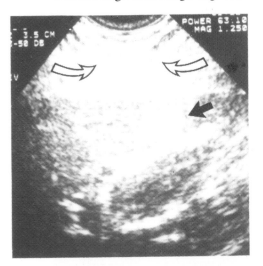

Tissue texture identified by hollow black arrows may be caused by beam interference rather than by reflectors in the tissue. This may represent acoustic speckle. (Source: Denis Gratton, Health Sciences Center, Winnipeg, Canada)

SLICE OR SECTION THICKNESS

An ultrasound beam has measurable thickness.

The imaging plane is assumed to be razor thin; therefore, the reflections produced by anatomic structures positioned slightly above or below the beam's main axis appear in the image.

In addition, hollow structures such as cysts may appear as filled-in, solid masses.

Extra echoes appear in the image. (Source: ATS Laboratories.)

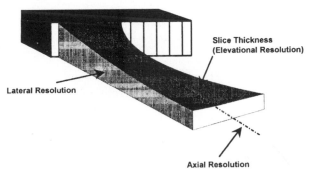

REFRACTION (See page 67.)

We assume that sound travels in a straight line.

However, waves can deflect when traveling from one media to another.

Therefore, reflections created after refraction appear on the screen in
improper locations.

The anatomic structure at position A is artifactually placed at B on the
image.

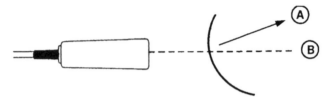

Generally, refraction artifact places a second copy, or duplicate, of the
anatomic structure on the image. The artifact replica (B) and the
correct anatomic reflection (A) are positioned side-by-side at almost
the same depth.

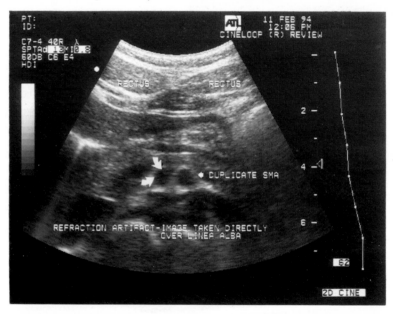

Refraction artifact of the superior mesenteric artery. The actual artery
(curved white arrows) is on the left, while the duplicate artifact is on the
right. (Source: Steven J. Bernhardt, Ochsner Clinics, New Orleans, Louisiana)

REVERBERATION (See page 91.)

Reverberations appear as multiple equally spaced reflections on the image.

When two or more strong reflectors lie in the path of a pulse, the sound
"ping-pongs" back and forth between the reflectors.

We assume that sound travels directly to the reflector and back to the
transducer.

It is easy to identify reverberations since they are *equally spaced* on the
display and have the appearance of the rungs of a ladder.

Reverberations often appear in small parts or water path scanners.

> Reverberation artifact places too many echoes on the image. The
> first and second reflections, closest to the transducer, are real.
> The remaining echoes, appearing at ever increasing depths, do
> not arise from anatomic structures.

COMET TAIL

Comet tail is a form of reverberation produced when two or more closely-
spaced, strong reflectors reside in a medium with a very high
propagation speed.

When a comet tail is produced, sound does not travel directly to a reflector
and back to the transducer; rather, it "ping pongs."

A strong linear echo appears at and extends deeper than the reflector.

A single, long, extra echo appears on the image that is not related to
anatomic structures.

Comet tail may result from reverberations that are closely spaced and
actually merge.

It may also arise from a small structure, such as a gas bubble, *resonating
or vibrating* after bombardment by the ultrasound pulse. The reflected
echo is long and appears as a linear echo extending downward from the
resonating structure. This is sometimes called **ring-down artifact**.

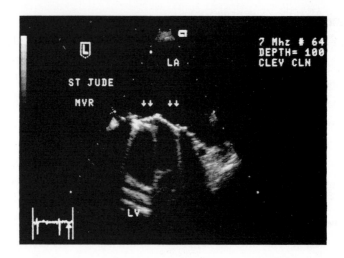

Comet tail and reverberation artifacts arising from the discs of an artificial, mechanical heart valve are displayed in this image. The white arrows identify the comet tail. The reverberations appear in lower portion of the image above the label "LV." (Source: Acuson)

MIRROR IMAGE

Sound can bounce off a strong reflector, such as the diaphragm, that is positioned in its path.

The structure acts as a mirror and redirects the pulse laterally toward a second reflector.

Ultrasound systems assume that sound travels directly to the second reflector and back to the transducer. Therefore, a second image of the reflector is incorrectly placed on the scan.

How to identify a mirror image:

- A duplicate of real anatomy appears on the image.
- The mirror image duplicate always appears deeper than the true anatomic structure.
- The anatomic structure acting as the "mirror" always lies on a direct line between the transducer and the artifact.

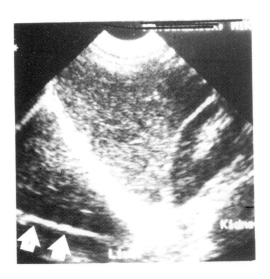

A longitudinal scan of the liver and kidney displays a mirror image artifact. The artifact identified by the white arrows was produced by sound waves reflecting off the strong diagonal structure and directed to the kidney/liver interface. (Source: Denis Gratton, Health Sciences Center, Winnipeg, Canada)

MULTIPATH

Multipath appears when the pulse travels by different paths to and from a reflector.

The pulse glances off a second structure on the way to or from the primary reflector.

The violated assumption is that a pulse travels directly to a structure and then back to the transducer.

The reflector is positioned at an incorrect depth. It is positioned deeper than the true anatomic structure.

Multipath is not explicitly seen on a scan; rather, it generally degrades image quality and diminishes longitudinal resolution.

SIDE LOBES

In reality, the acoustic pulse produced by a transducer may not have the ideal geometry of an hourglass. In some cases, significant acoustic energy is emitted by the transducer in a direction different from that of the main axis of the sound beam.

Side lobes are produced by **single crystal transducers**. The acoustic energy in the lobes is less than that in the primary beam.

Nonetheless, when a very bright reflector lies in the path of a side lobe, a reflection of sufficient strength to appear on the image may be generated.

The violated assumption is that reflections are produced by structures located only along the main axis of the sound beam.

Reflections received by the transducer from these artifactual, off-axis sound pulses appear on the image as if they reside in the main beam.

A duplicate of the true reflector appears on the image at its correct depth, but it is positioned laterally from the true anatomy.

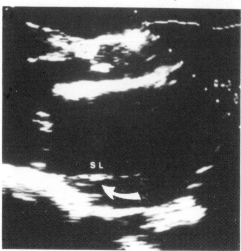

The horizontal echo identified by the white arrow and labeled "SL" is a side lobe artifact. It does not represent anatomy; rather, it is produced by the strong horizontal reflector that appears to its left. (Source: Leonard Pechacek, Houston, TX)

GRATING LOBES

Array transducers produce off-axis acoustic waves as a result of the regular, systematic spacing of active elements. The transducer may emit acoustic energy in a direction other than that of the beam's main axis.

Reflectors in the path of the grating lobe may appear in improper locations on the image. This artifact is not commonly seen because of a corrective process, called sub-dicing, performed during the transducer manufacturing process.

The acoustic energy in the grating lobe is less than that in the main beam. Nonetheless, when a very bright reflector lies in the path of a grating lobe, a reflection of sufficient strength to appear on the image may be generated.

The violated assumption is that reflections are produced by structures
located along the main axis of the sound beam.

Reflections received by the transducer from these artifactual, off-axis
sound pulses appear on the image as if they reside in the main beam.

A duplicate of the true reflector appears on the image at its correct depth,
but is it positioned laterally from the true anatomy.

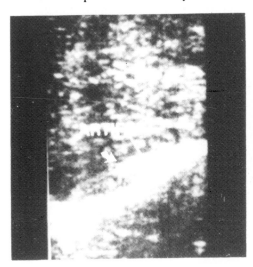

Transverse scan of the liver,
produced with an old technology
linear array scanner, has a
grating lobe artifact identified
by the white arrows. (Source:
Denis Gratton, Health Sciences
Center, Winnipeg, Canada)

SHADOWING

Shadows are the erroneous absence or reduced brightness of reflectors on
an image caused by the weakening of a sound beam.

There is *too little information* on a scan that exhibits shadowing.

Two causes of shadowing:

- Abnormally high attenuation.
- Refraction.

Shadows may be helpful in arriving at a clinical diagnosis.

Shadow artifacts are the *same color as the background* of the image.
Thus, shadows appear black on a white-on-black image. Shadows
appear white on a black-on-white image.

The violated assumption is that the strength of a reflection is directly
related to the scattering strength of the object that is imaged.

Shadowing by attenuation: Absent reflectors on an image when ultrasound pulses travel through structures with abnormally high attenuation (such as bone). The acoustic energy and reflecting echoes from structures deeper than the attenuator are greatly diminished.

Shadowing by refraction: Refraction causes an ultrasound beam to change direction and *spread out* (diverge or defocus). Reflections from a defocused beam have reduced strength and fail to appear on the image. Therefore, reflectors in the body are absent on the image. Areas deeper than the refractor are abnormally anechoic.

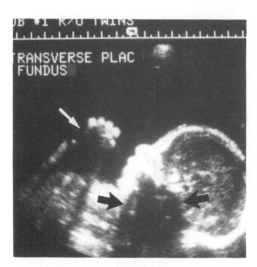

Shadowing behind the mandible (solid black arrows) and the fist (solid white arrows) of the fetus results from attenuation of the sound beam at the bone/tissue boundary. (Source: Acuson)

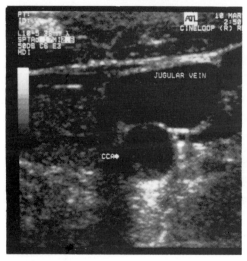

Shadowing along the edge of the jugular vein. The shadow artifact, directly in line with the label "CCA," may result from refraction and defocusing of the sound beam as the beam propagates along the edge of the jugular vein. (Source: Advanced Technology Laboratories)

ENHANCEMENT

Enhancement is the term for overly bright reflectors that appear on the
   image when the attenuation of the sound pulse is less than anticipated.
The number of reflectors on the image is correct, but some are too bright.
Two causes of enhancement:
   ♦ Abnormally low attenuation.
   ♦ Focusing.
Enhancement may assist in making a clinical diagnosis.
Enhancement artifact appears as the color *opposite the image's
   background*. Thus, enhancement appears white on a white-on-black
   image and black on a black-on-white image.
The violated assumption is that the strength of a reflection is directly
   related to the scattering strength of the object that is imaged.
Enhancement by attenuation: This occurs when sound travels through a
   medium with an attenuation rate that is lower than that of surrounding
   tissue. Reflectors deeper than the weak attenuator are abnormally
   bright in comparison to neighboring tissues.
Enhancement by focusing: Reflections from a highly focused pulse may
   have abnormally strong intensities in the focal zone and may appear
   abnormally bright. When a beam is strongly focused, the concentration
   of energy within the beam is increased. Therefore, certain structures
   appear brighter than others as a result of their position near the focus.
   An entire horizontal region of tissue at the focal depth may appear
   overly bright. This particular form of enhancement is called **banding**.

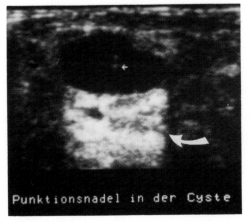

Punktionsnadel in der Cyste

Enhancement of tissues (white
arrow) deeper than the breast
cyst are displayed. The
attenuation of the sound through
the cyst is less than that of the
surrounding tissues and results
in this abnormal brightness.
(Source: Acuson)

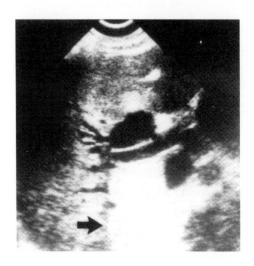

Transverse scan of the liver showing enhancement (black arrow) at depths greater than those of the grossly dilated common hepatic bile duct. (Source: Denis Gratton, Health Sciences Center, Winnipeg, Canada)

PROPAGATION SPEED ERRORS

Reflectors are placed in improper positions on an image when sound travels at a speed other than 1,540 m/s. This occurs when sound travels through media at speeds unequal to those in soft tissue.

The reflectors in the scan appear in *correct number* but at *improper depths*. Measurements made on these images are inaccurate.

When the medium's speed is *greater* than that in soft tissue:
- Sound travels faster than the ultrasound system expects.
- Pulses return from trip in body - fast, Fast, FAST!!
- The go-return time is too short.
- Machine assumes reflectors are close to the transducer.
- Reflectors are too shallow.
- All distances are underestimated (reported number is too small).

When the speed of sound in a tissue is 10% greater than that in soft tissue, the go-return time is 10% less than normal. As a result, all distances will be underestimated by 10%.

When the medium's speed is *less* than that in soft tissue:

- Sound travels slower than the ultrasound system expects.
- Pulses return from trip in body - slow, Slow, SLOW!!
- The go-return time is too long.
- Machine assumes reflectors are far from transducer.
- Reflectors are placed too deep.
- All distances are overestimated (reported number is too large).

When the speed of sound in a tissue is 10% less than that in soft tissue, the go-return time is 10% longer than normal. As a result, all distances will be overreported by 10%.

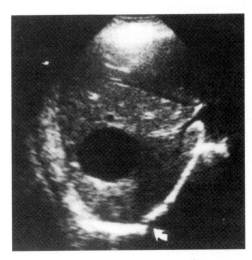

Scan of the liver. Note that the diaphragm directly below the circular mass within the liver appears to have a discontinuity (white arrow). This propagation speed artifact, the apparent interruption of the diaphragm, may result from the difference of the speed of sound between the mass and surrounding tissues. (Source: Denis Gratton, Health Sciences Center, Winnipeg, Canada)

ARTIFACTS - SUMMARY

Anatomic reflectors absent on image:
- Shadowing
- Lateral resolution
- Axial resolution

Anatomic reflectors appear with abnormal brightness:
- Enhancement (too bright)
- Banding (too bright)
- Shadowing (too dark)

Single anatomic reflector appearing multiple times on image:

Artifact positioned *directly below* true anatomy
- Comet tail
- Ring down
- Reverberations
- Multipath

Artifact displaced *to the side* of the true anatomy
- Refraction
- Side lobe
- Grating lobe

Artifact positioned *deeper and displaced to the side* of the true anatomy
- Mirror image

Anatomic structures appearing at incorrect depth:
- Speed errors

Anatomic structures appearing in the incorrect imaging plane:
- Slice or section thickness

Anatomic structures not corresponding to echoes on the image:
- Acoustic speckle

QUESTIONS - ARTIFACTS

1. When creating an ultrasound image, all of the following assumptions are made EXCEPT:
   a) Sound travels in a straight line.
   b) Sound travels at 1.54 km/sec.
   c) Reflections only arise from structures in the pulse's main beam.
   d) The sound beam is extremely thin.
   e) All structures create reflections of equal magnitude.

2. Which statement regarding axial resolution artifact is incorrect?
   a) Axial resolution artifact is related to beam diameter.
   b) Higher quality images are associated with smaller numbers.
   c) Axial resolution is reported in units of millimeters.
   d) Numerically, axial resolution equals one-half the pulse length.
   e) Too few reflectors appear on the image.

3. An ultrasound pulse has a width of 4 mm, a length of 2 mm, and is produced by a transducer 3,000 times per second. What is the best estimate of the system's radial resolution?
   a) 4 mm   b) 2 mm   c) 1 mm   d) 1,500 Hz

4. Which of the following statements is true about lateral resolution?
   a) It is also called angular, transverse, and depth resolution.
   b) It is reported in units of time.
   c) It is the same at all depths.
   d) It can be improved by focusing.

5. All of the following artifacts result in the placement of too many echoes on the image EXCEPT:
   a) shadowing   b) reverberation
   c) mirror image   d) grating lobes

6. Which artifact produces an image with an incorrect number of echoes?
   a) propagation speed error b) multipath
   c) enhancement   d) side lobes

7. Which artifact is unrelated to the shape of an ultrasound pulse?

      a) lateral resolution   b) depth resolution

      c) slice thickness   d) refraction

8. Two reflections, one true and one artifact, are displayed on an ultrasound image. In the body, only one anatomic structure is present. The correct reflection and the artifact are found side-by-side. What is the most likely cause of this artifact?

      a) mirror image   b) grating lobe

      c) enhancement   d) multipath

9. Two distinct reflections are observed on an image, but they actually arise from a single anatomic structure. The artifact is positioned deeper than the correct reflection. What is the most likely cause of this artifact?

      a) grating lobe   b) side lobe

      c) refraction   d) mirror image

10. Which one of these artifacts does not result from multiple reflections?

      a) comet tail   b) reverberations

      c) ring down   d) enhancement

ANSWERS - ARTIFACTS

1 - e: The magnitude of a reflection is related to the scattering strength of the anatomic structure that creates it. For example, reflections from bone are usually strong, whereas reflections from soft tissue are normally weak.

2 - a: Axial resolution is not related to beam diameter. It is related to the length of the pulse.

3 - c: Radial resolution is approximately equal to one-half of the pulse's length. The length is 2 mm. Therefore, the radial resolution is 1 mm.

4 - d: Lateral resolution, measured in millimeters, depends upon the width of the ultrasound pulse. Focusing narrows the beam width, thereby improving lateral resolution. Choice a is incorrect because depth resolution is not synonymous with lateral resolution.

5 - a: Shadowing artifact is the incorrect absence of reflectors on the image when a sound pulse is attenuated.

6 - d: Side lobe artifact places a second echo on an image from a single structure. The other choices create images with the correct number of echoes.

7 - d: Refraction artifact is not related to the shape of a sound pulse. Refraction artifact depends upon a sound wave striking a boundary with oblique incidence. The propagation speeds of the tissues on either side of the boundary must be dissimilar.

8 - b: Grating lobe artifact creates a second copy of an anatomic reflector that is positioned alongside the true reflection.

9 - d: A mirror image artifact is positioned deeper than the structure that created it. The other choices generally create artifacts that are located at the same depth as the true reflector.

10 - d: Enhancement is an artifact related to abnormal brightness. All of the other choices are artifacts associated with multiple reflections.

QUALITY ASSURANCE

QUALITY ASSURANCE (quality control) is the routine, periodic evaluation of an ultrasound system to guarantee optimal image quality. It is impossible to specify the appropriate time interval between quality assurance evaluations: this is dependent on the clinical circumstances.

> Q/A is performed routinely.

The four requirements for quality assurance programs:
- A variety of tests to evaluate the system's components.
- Repairs.
- Preventative maintenance.
- Record keeping.

> An ultrasound system is only as strong as its weakest component.

Goals:
- Proper operation of equipment.
- Detection of gradual changes.
- Minimization of downtime.
- Reduction of non-diagnostic exams.
- Reduction of repeat scans.

Medical/Legal Aspects: Quality assurance is a necessity for every laboratory.

Note: Various aspects of instrumentation should be *routinely evaluated* to ensure consistency of its performance. It is important to validate the reliability of the images produced and the measurements made on each ultrasound system with each transducer.

Numerous devices enable the sonographer to perform quality assurance.

Four of these devices will be presented and discussed:
- AIUM 100 mm test object.
- Tissue equivalent phantom.
- Doppler phantom.
- Beam profile/slice thickness phantom.

AIUM 100 MM TEST OBJECT

This test object is commercially available and contains an array of strategically located pins or nylon fibers. The accuracy and performance characteristics of a system may be evaluated by scanning the object. It has a propagation speed similar to that of soft tissue: 1,540 m/s. On the other hand, it *does not* have attenuation properties of soft tissue. US systems can be evaluated for depth accuracy, dead zone, linearity, axial and lateral resolution, and position registration. This device is important in evaluating B-scanners and A-mode systems. Its lack of similarity to soft tissue limits the evaluation of attenuation or gray scale. (Source: Nuclear Associates)

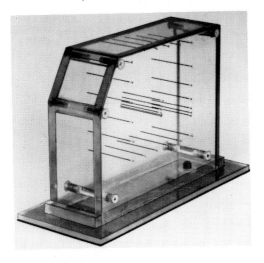

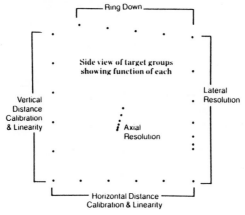

TISSUE EQUIVALENT PHANTOM

This commercially available object contains a medium that *mimics soft tissue.*
It has an acoustic velocity of 1,540 m/s and an attenuation coefficient
similar to that of soft tissue. Imbedded in the phantom are strategically
located pins or nylon fibers (similar to the AIUM 100 mm test object)
and structures that mimic hollow cysts and solid masses.

This is a useful device in the evaluation of tissue texture images. A tissue-
mimicking device is appropriate for assessing image quality from
modern day, real-time scanners. These phantoms are especially useful
when evaluating multi focus and adjustable-focus phased array
transducers. (Source: Gammex RMI)

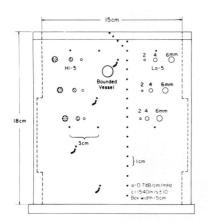 

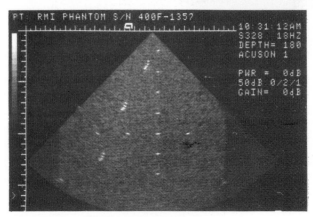

Tissue equivalent phantom scan assessing quality assurance. (Source: Acuson)

DOPPLER PHANTOM

This phantom is a commercially available system containing simulated vessels positioned at a variety of angles to the imaging surface. A pumping system forces echogenic fluid through the vessels at known velocities, pulse rates, and durations. In addition, a constriction, or stenosis, is present in one of the vessels. With these features, the phantom can be used to assess the accuracy of pulsed, continuous wave, and color systems.

An alternative design of Doppler phantoms includes a string that moves while it is immersed in a water bath. Echoes from the **vibrating string** imitate reflections from red blood cells. (Source: Gammex RMI)

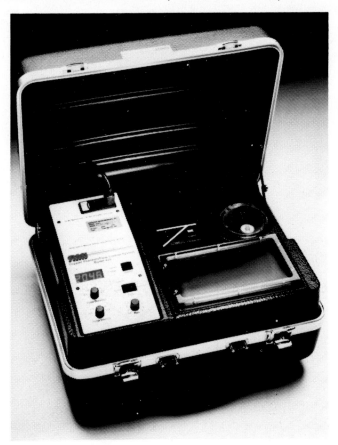

BEAM PROFILE/SLICE THICKNESS PHANTOM

This commercially available object contains a diffuse scattering plane at an
angle to the incident sound beam. The phantom's medium mimics soft
tissue. The slice thickness is a measure of the beam geometry in the
plane that is perpendicular to the imaging plane. This important
measurement is not obtained with other more traditional test objects.

Increased slice thickness will diminish spatial resolution and reduce the ability
to distinguish small, low-contrast lesions. When the US beam is overly
thick, cystic structures may appear filled-in. (See slice thickness artifact
on page 201.) (Source: ATS Laboratories)

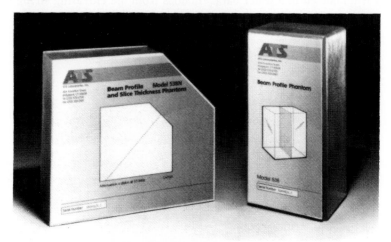

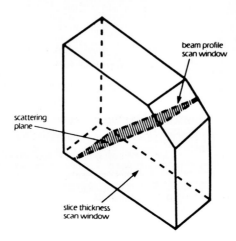

PERFORMANCE MEASUREMENTS

Each transducer should be evaluated by obtaining the following measurements. Although these measurements appear separate and distinct from each other, a single image obtained while scanning a phantom may provide quality assurance data for multiple performance criteria.

Minimum sensitivity is the smallest gain required to visualize a pin located deep in the test object. It is determined by setting the TGC flat and increasing the gain from its lowest value to the level where a particular pin or nylon string is observed. For all future evaluations, the same pin in the test object should be used.

By using the tissue equivalent phantom, when the system's controls are set to a particular level, tissue-like texture should be observed at the same depth.

Normal sensitivity is the gain setting required to visualize all of the pins in a test object. Normal sensitivity is, of course, found at a higher gain than minimum sensitivity. All other performance measurements are made at this gain setting.

When a tissue phantom is used, normal sensitivity establishes the maximum depth of tissue texture. With constant instrument settings, this depth should not change between evaluations.

Registration accuracy is the ability of a B-scanner to place echoes in proper positions while imaging from different orientations. The AIUM test object is scanned from the top and from both sides. Each pin should generate a star-like pattern when the B-scanner is properly registered. When misregistration occurs, a single pin imaged from different orientations will produce numerous, disassociated echoes.

Range accuracy or vertical depth calibration allows for assessment of the machine's ability to accurately position echoes. Calibration is accomplished with pins positioned at ever-increasing depths. If an error exists, it may be due to machine malfunction, or because the speed of sound in the phantom is not exactly 1.54 km/sec. With this technique the accuracy of the US system's electronic ruler (digital calipers) can be assessed.

Horizontal calibration is similar to vertical calibration, except that we evaluate the system's ability to correctly position reflectors that are located side by side.

Dead zone is the region close to the transducer that cannot be imaged accurately. It is determined by scanning the series of pins located near the top of the AIUM test object. Dead zone results from the time delay that elapses for the active element and the receiver electronics to switch from the transmit to receive mode.

An acoustic standoff can be used to prevent the dead zone from interfering with clinically important structures.

Focal zone is the region along the US beam where the intensity is the highest and the width of the beam is the narrowest (finest lateral resolution). Pins at ever-increasing depths are

Acoustic standoff.
(Source: Nuclear Associates)

scanned, and the width of the reflections on the image are measured. The location of the narrowest reflection is the focus.

A **beam profiler** can also be used to assess the focal zone with the use of an A-mode display format. This evaluation is especially important for those systems with sonographer-controlled focusing, such as phased array and annular transducers.

Longitudinal resolution is the smallest recorded distance at which pins are displayed as separate echoes when the objects are positioned one in front of the other. By using the set of pins that are positioned at ever-increasing depths, the smallest distance at which two pins are displayed as two separate echoes can be recorded. This distance is the system's longitudinal resolution. (See page 90.)

Lateral resolution defines the ability of an ultrasound system to accurately image objects that are side-by-side. The minimum distance at which two closely spaced, side-by-side rods are displayed as two separate images can be determined. This is the lateral resolution *at that depth*. Recall that LATA resolution varies with beam diameter. Therefore, the sonographer should measure LATA resolution at a variety of depths. (See page 92.)

Compensation operation or uniformity describes the ability of a system to display similar reflectors, such as tissue texture, as identical echoes on the screen of the CRT regardless of their position. To evaluate uniformity, a tissue phantom is the preferred test object.

The AIUM test object may also be used to evaluate uniformity. While scanning a group of pins at ever-increasing depths, the echoes are displayed with TGC off and then with TGC adjusted. With the TGC off, the echoes should be displayed with reduced amplitude as depth increases. With the TGC on, all of the echoes should have the same amplitude, regardless of the depth from which they return.

Mock cysts and tumors - By using the tissue equivalent phantom, inspect the dimensions and shapes of the mock cysts and solid masses. Also, note texture and fill-in of both hollow and solid masses.

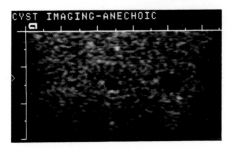 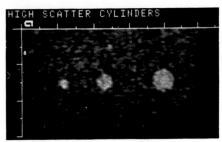

Scan of tissue equivalent phantom's mock cysts and tumors. (Source: Gammex RMI)

Display and gray scale dynamic range - With any group of pins, adjusting the gain of the system should produce changes in the gray scale or brightness of the echoes. By monitoring the system during power-up, the accuracy of the initializing sequence can be determined. One should compare the images appearing on the screen of the machine with those of other accessory devices.

MEASURING THE OUTPUT OF ULTRASOUND MACHINES

Although it is clinically important, the sonographer rarely measures the output of an ultrasound probe. Typically, this task is left to the equipment manufacturer, engineer, and scientist. The devices described below are essential for optimal design of an ultrasound system. These devices make it possible to conduct meaningful bioeffects research.

A hydrophone is a small probe, less than 1 mm in diameter, with a piezoelectric crystal at its end (a small ultrasound transducer). On occasion, a PZT crystal is attached to the tip of a small needle that is inserted into the ultrasound beam. The dimensions of the hydrophone are so small that the sound beam is essentially undisturbed. Hydrophones in the form of a thin, plastic membrane are used to measure acoustic pressure. The output of the device is transferred to an oscilloscope, which displays acoustic signals as received by the PZT. (Source: Nuclear Associates)

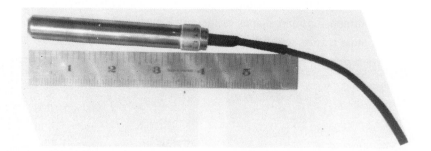

Acousto-optics are based on the interaction between two types of waves: sound and light. This method allows quantitation of sound beam's characteristics and the shape of the beam. A shadowing system, called a **Schlieren**, uses this principle to measure beam profiles.

Radiation force balance - A sound wave striking an object exerts a small, but measurable force. A small scale, called a radiation force balance, is placed in the ultrasound field to measure the force applied by the beam. Other force-measuring systems use a balance or a float, acting as a tiny scale, to measure the acoustic power in the beam.

A calorimeter is a "transducer" that converts acoustic energy into heat. When the ultrasound beam is aimed inside the calorimeter, the acoustic energy is completely absorbed and converted into heat. When the temperature increase is measured along with the time required to obtain the heat, the total power of the ultrasound beam can be calculated.

A thermocouple is a small device imbedded in an absorbing material. The sound is absorbed (converted into heat). The thermocouple measures changes in temperature. The thermocouple uses principles similar to those employed by a calorimeter, namely the conversion of acoustic energy to heat energy. However, the ultrasound intensity at a specific location in space is measured by a thermocouple.

Crystals - When cholesteric liquid crystals (or starch/iodine blue) are struck by different ultrasound intensities, their colors change. The shape and colors of the crystals provide insight into the shape and strength of the beam.

BIOLOGIC EFFECTS AND SAFETY

The first commandment regarding clinical ultrasound is that the benefits to the patient must outweigh the risks of the exam.

> Clinicians must balance the *risks* and *benefits* of any procedure.

Dosimetry is the science of identifying and measuring the characteristics of an ultrasound field that are relevant to its potential for producing biological effects.

What are these characteristics? WE AREN'T SURE!

We know that very high intensities of US cause damage to biologic tissues. But, there are no known cases of diagnostic imaging at standard intensities resulting in biological effects and tissue injury. One of the most common applications of ultrasound is in the area of fetal medicine. Even after years of clinical use, there is no confirmation of harm resulting from its intended use.

References: *Bioeffects Considerations for the Safety of Diagnostic Ultrasound.* J. of Ultrasound in Medicine. Vol. 7 (suppl), 1-38, Sept. 1988. *Bioeffects & Safety of Diagnostic Ultrasound*, American Institute of Ultrasound in Medicine, 11200 Rockville Pike, Suite 205, Rockville, MD 20852-3139, 1993.

Note: AIUM statements and conclusions found in this section were published in one of the above references.

TECHNIQUES TO STUDY BIOEFFECTS

MECHANISTIC APPROACH:
- Propose a specific physical mechanism that has the potential to produce bioeffects.
- Perform a theoretical analysis to estimate the scope of bioeffects at various exposure levels.
- Identify "cause-effect" relationship.

EMPIRICAL APPROACH:
- ◆ Acquire or review data from patients or animals.
- ◆ Correlate exposure and effects.
- ◆ Identify an "exposure-response" relationship.

STRENGTHS (+) and WEAKNESSES (-) OF THE METHODS

Mechanistic:
- ◆ A broad variety of exposures can be studied (+).
- ◆ Are the assumptions valid? (-)
- ◆ Are other mechanisms involved? (-)

Empirical:
- ◆ There is no need to understand the mechanism. (+ and -)
- ◆ The biological significance of the mechanism is obvious. (+)
- ◆ Different species may produce different results. (-)
- ◆ Statistical problems. (-)

> A firm conclusion regarding bioeffects is justified when both
> mechanistic and empirical results are consistent.

MECHANISMS OF BIOEFFECTS

Two important mechanisms of bioeffects, **thermal** and **cavitation**, necessitate
significant discussion.

Additionally, scientists have proposed that **physical vibration** of biologic
media may induce bioeffects. Neither empirical nor mechanistic
evaluations of the vibration mechanism has shown evidence of
bioeffects from pure and direct mechanical vibration.

THERMAL MECHANISM

The **THERMAL MECHANISM** proposes that ultrasound may induce
  bioeffects by elevating tissue temperature.

Rationale for studying the thermal mechanism:

- As sound propagates in the body, acoustic energy is converted into heat.
  Tissue temperature elevation may occur via absorption.
- Body temperature is regulated at 37° C core. Life processes do not
  function normally at other temperatures.

Alternatively, some ultrasound experts believe that it is unnecessary to
  investigate a thermal mechanism because it is quite common that a
  2^O increase in body temperature may occur during fever, exercise,
  and sunbathing without any adverse effects.

Conclusion: Investigations of heat-induced bioeffects resulting from exposure
  to ultrasound should be undertaken.

Thermal Mechanism - Empirical Data

Serious tissue damage occurs from prolonged elevation of body temperature by
  2.5^O C or 4.5^O F.

A 2^O to 4° degree rise in testicular temperature can cause infertility.

A combination of temperature and exposure time determine the likelihood of
  harmful bioeffects. The higher the temperature, the shorter the
  exposure time needed to produce harmful effects.

No confirmed bioeffects have been reported for temperature elevations of up to
  2^O C above normal for exposures of less than 50 hours.

Maximal heating is related to the beam's SPTA intensity.

Fetal tissues appear less tolerant of temperature elevations than adult tissues.

A greater amount of acoustic energy is absorbed by bone than by soft tissue.
  Therefore, circumstances wherein ultrasound strikes fetal bone
  deserve special attention.

Thermal Mechanism - Mechanistic Data

Despite the following, the theoretical models appear to correlate with
experimental data:

- The ultrasound beam structure is complex.
- Diagnostic equipment is diverse.
- Tissue characteristics are different.

CONCLUSIONS REGARDING A THERMAL BIOEFFECTS MECHANISM

Approved by AIUM, October 1987

1. A thermal criterion is one reasonable approach to specifying potentially hazardous exposures for diagnostic ultrasound.

2. Based solely on a thermal criterion, a diagnostic exposure that produces a maximum temperature rise of 1^O C above normal physiological levels may be used in clinical examinations without reservation.

3. An in situ temperature rise to or above 41° C is considered hazardous in fetal exposures; the longer this temperature elevation is maintained, the greater is the likelihood for damage to occur.

4. Analytical models of ultrasonically induced heating have been applied successfully to in vivo mammalian situations. In those clinical situations where local tissue temperatures are not measured, estimates of temperature elevations can be made by employing such analytical models.

5. Calculations of ultrasonically induced temperature elevation, based on a simplified tissue model and a simplified model of stationary beams, suggest the following: For examinations in fetal soft tissues with typical perfusion rates, employing center frequencies between 2 and 10 MHz and beam widths less than 11 wavelengths, the computed temperature rise will not be significantly above 1^O C if the in situ SATA intensity does not exceed 200 mW/cm^2. If the beam width does not exceed eight wavelengths the corresponding intensity is 300 mW/cm^2. However, if the same beam impinges on fetal bone, the local temperature rise may be higher.

CAVITATION

CAVITATION is the interaction of sound waves with microscopic, stabilized, gas bodies (called gaseous nuclei) in the tissues.

There is great uncertainty regarding the nature of these tiny gas bubbles with regard to their location, size, chemical composition, conditions under which they occur, and materials in which they exist.

The existence of gas bubbles in the body has been confirmed from lithotripter research. Lithotripsy is a nonsurgical procedure that utilizes shock waves to smash kidney stones into small particles.

Unlike thermal bioeffect, which is related to the temporal average intensity of a sound beam, the cavitation bioeffect is related to the temporal peak intensity of the beam.

Two forms of cavitation occur: **stable** and **transient**.

STABLE CAVITATION

In stable cavitation, bubbles tend to *oscillate* when exposed to acoustic waves of small amplitude. They repetitively swell and contract. Bubbles a few micrometers in diameter double in size.

Bubbles intercept and absorb much of the acoustic energy. Shear stresses and microstreaming are produced in surrounding fluid.

TRANSIENT CAVITATION

In transient cavitation (also called normal or inertial cavitation), bubbles expand and collapse violently, causing a microscopic explosion. In other words, the bubble *bursts*. Bursting depends upon the pressure of the ultrasound pulses.

Pressure units: MPa, mega pascals.

Note: The pressure threshold for transient cavitation is only 10% greater than that required for stable cavitation.

Transient cavitation produces highly localized violent effects:
- Enormous pressures, shock waves, mechanical stress.
- Potential for colossal temperatures.

Cavitation is not considered a clinically significant risk to patients because if harmful effects do occur, they are found only in a few cells.

CONCLUSIONS REGARDING CAVITATION
Approved by the AIUM, October 1992

1. Acoustic cavitation may occur with short pulses and has the potential for producing deleterious biological effects. The temporal peak outputs of some currently available diagnostic ultrasound devices can exceed the threshold for cavitation in vitro and can generate levels that produce extravasation of blood cells in the lungs of laboratory animals.

2. A number, called the Mechanical Index, has been developed to predict the likelihood of cavitation induced bioeffects.

3. No confirmed biologically significant adverse effects have been reported in mammalian tissues that do not contain well-defined gas bodies.

EPIDEMIOLOGY

EPIDEMIOLOGY is a branch of medicine dealing with the prevalence of disease. It is empirical and utilizes clinical surveys.

Most epidemiologic bioeffects studies deal with in utero fetal exposures to ultrasound.

IN UTERO EXPOSURE
Concerns for ultrasound bioeffects exist because :
- Half of all pregnant women in the United States are scanned. Thus, there are major public health implications.
- Ultrasound is used often during normal pregnancies, thereby skewing the risk-benefit ratio.
- Harmful effects have the potential to affect the fetus for decades.

Epidemiological studies have evaluated these factors: birthweight, structural anomalies, neurological development of children, cancer, and hearing.

LIMITATIONS OF EPIDEMIOLOGIC STUDIES

- Studies are often retrospective. (Information is collected from old medical records.)

- Ambiguities exist in the data, such as justification for the exam, gestational age, number of exams, exposure, and mode.

- Risk factors other than exposure to ultrasound may precipitate a bad outcome in the fetus. These include maternal age, poor nutrition, smoking, alcohol, and drug abuse.

STATISTICAL CONSIDERATIONS

- The smaller the bioeffect, the harder it is to detect.

- Thus, before valid conclusions regarding bioeffects can be made, large numbers of patients must be included in the research populations.

STATISTICAL POWER answers the question: "How many patients are required to detect a harmful effect in a statistically valid manner?"

Example: If a 5% event rate occurs naturally and ultrasound exposure increases this rate to 5.5%, then 5,200 patients would be required to confirm this increase in a statistically valid manner.

CONCLUSIONS REGARDING EPIDEMIOLOGY
Approved by the AIUM, October 1987

1. Widespread clinical use over 25 years has not established any adverse effect arising from exposure to diagnostic ultrasound.

2. Randomized clinical studies are the most rigorous method for assessing potential adverse effects of diagnostic ultrasound. Studies using this methodology show no evidence of an effect on birthweight in humans.

3. Other epidemiologic studies have shown no causal association of diagnostic ultrasound with any of the adverse fetal outcomes studied.

IN VIVO AND IN VITRO STUDIES

IN VIVO BIOEFFECT INVESTIGATIONS

In vivo means "observed in living tissues."

The following are recent conclusions of in vivo bioeffects investigations:

> When compared with unfocused beams, focused beams require higher intensities to produce bioeffects. This occurs because smaller beam area means less thermal buildup and less interaction with cavitation nuclei.

Note: An unfocused ultrasound beam causes a higher temperature elevation than a focused ultrasound beam at the same intensity.

> It has been proved that, compared with a broad unfocused beam, highly focused ultrasound is much less likely to cause bioeffects.

Maximum intensities (SPTA): 100 mW/cm^2 - unfocused.
1 W/cm^2 - focused.

CONCLUSIONS REGARDING IN VIVO MAMMALIAN BIOEFFECTS
Approved by the AIUM, October 1992

In the low megahertz frequency range there have been (as of this date) no independently confirmed significant thermal biological effects in mammalian tissues exposed in vivo to unfocused ultrasound with intensities below 100 mW/cm^2, or to focused ultrasound with intensities below 1 W/cm^2 SPTA.

IN VITRO BIOEFFECT INVESTIGATIONS

In vitro means "observed in test-tubes" in an experimentally controlled environment.

Advantage of in vitro studies: Careful measurements can be made under rigorous experimental conditions.

Disadvantage of in vitro studies: Results may not pertain to in vivo bioeffects. Nonetheless, bioeffects observed in vitro must be considered real.

AIUM STATEMENT ON IN VITRO BIOLOGICAL EFFECTS
Approved by the AIUM, March 1988

It is difficult to evaluate reports of ultrasonically induced in vitro biological effects with respect to their clinical significance. The predominant physical and biological interactions and mechanisms involved in an in vitro effect may not pertain to the in vivo situation. Nevertheless, an in vitro effect must be regarded as a real biological effect.

Results from in vitro experiments suggest new endpoints and serve as a basis for design of in vivo experiments. In vitro studies provide the capability to control experimental variables and thus offer a means to explore and evaluate specific mechanisms. Although they may have limited applicability to in vivo biological effects, such studies can disclose fundamental intercellular or intracellular interactions.

While it is valid for authors to place their results in context and to suggest further relevant investigations, reports of in vitro studies which claim direct clinical significance should be viewed with caution.

AIUM STATEMENT ON CLINICAL SAFETY
Approved March 1988, Reaffirmed 1992

Diagnostic ultrasound has been in use since the late 1950s. Given its known benefits and recognized efficacy for medical diagnosis, including use during human pregnancy, the American Institute of Ultrasound in Medicine herein addresses the clinical safety of such use:

No confirmed biological effects on patients or instrument operators caused by exposure at intensities typical of present diagnostic ultrasound instruments have ever been reported. Although the possibility exists that such biological effects may be identified in the future, current data indicate that the benefits to patients of the prudent use of diagnostic ultrasound outweigh the risks, if any, that may be present.

AIUM STATEMENT ON SAFETY IN TRAINING AND RESEARCH
Approved March 1988

Diagnostic ultrasound has been in use since the late 1950s. No confirmed adverse biological effects on patients resulting from this usage have ever been reported. Although no hazard has been identified that would preclude the prudent and conservative use of diagnostic ultrasound in education and research, experience from normal diagnostic practice may or may not be relevant to extended exposure times and altered exposure conditions. It is therefore considered appropriate to make the following recommendation:

In those special situations in which examinations are to be carried out for purposes other than direct medical benefit to the individual being examined, the subject should be informed of the anticipated exposure conditions, and of how these compare with conditions for normal diagnostic practice.

When there is no direct medical benefit to a person undergoing an ultrasound exam (e.g., training or research), it is necessary to *educate* the person regarding the risks of the procedure and obtain his or her *informed consent*.

> The AIUM suggests the following:
> - Do not perform studies without reason.
> - Do not prolong studies without reason.
> - Use the minimum output power and maximum amplification to optimize image quality.

ELECTRICAL AND MECHANICAL HAZARDS

Several different instruments, as well as the individual components of the ultrasound system, may be linked to a patient at any given time. Precautions such as proper electrical grounding should always be taken to avoid electrical hazard. Instruments should be routinely checked for proper condition.

However, ultrasound systems present no special electrical safety hazards. Mechanically, the machine should be inspected to assure proper physical status. Since the transducer is in direct contact with the patient, it may be considered the component most likely to pose a threat, albeit small, to a patient.

In summary:

be prudent, use your head, be careful, be judicious.

BE AN ULTRASOUND PROFESSIONAL

QUESTIONS - BIOEFFECTS AND SAFETY

1. A researcher studying bioeffects of ultrasound reports that obstetrical scanning may be harmful to a particular group of patients. What should be the response of the medical community?
   a) Perform the exams on all patients when the benefits are outweighed by the risks.
   b) Stop all diagnostic ultrasound exams.
   c) Ignore the report.
   d) Perform no exams on this group of patients.
   e) Perform exams on all patients when the benefits outweigh the risks.

2. The study of the characteristics, attributes, and quantities of a substance that induce bioeffects is called _____.

3. True or False? There are no bioeffects associated with ultrasound.

4. True or False? There are no bioeffects associated with ultrasound with characteristics typical of those in diagnostic medicine.

5. True or False? There are no harmful bioeffects associated with ultrasound with characteristics typical of those in diagnostic medicine.

6. Which of the following is not associated with the mechanistic approach to the study of bioeffects?
   a) Identifying a cause and effect relationship.
   b) Proposing a specific means that could produce a bioeffect.
   c) Studying charts of patients who have been exposed to ultrasound.
   d) Analyzing the mechanism using theoretical methods.
   e) Arriving at a valid scientific conclusion.

7. Veterans of "Operation Desert Storm" suspect that they have been exposed to chemical warfare agents. The Veterans Administration Hospital surveyed these soldiers to identify the presence of symptoms and their prevalence. What form of medical investigation is this?

8. Name the two important mechanisms likely to induce bioeffects.

9. Which of these statements has allowed scientists to conclude that diagnostic ultrasound generally does not injure tissues via temperature elevation?
   a) Death resulting from an ultrasound exam has never been reported.
   b) Mechanistic data indicate that diagnostic ultrasound is safe.
   c) Humans often experience temperature elevations that are unrelated to exposure to ultrasound.
   d) Conclusions from empirical data and mechanistic data are consistent, i.e., diagnostic ultrasound is unlikely to cause thermally induced bioeffects.
   e) Patients do not complain of burning or sweating during US exams.

10. True or False? Cavitation describes the interaction between sound waves and small gas bubbles that inhabit the tissues.

11. True or False? There are two forms of cavitation: inertial and normal.

12. True or False? Normal, transient or inertial cavitation describes the bursting of microbubbles.

13. True or False? Stable cavitation describes the rhythmical swelling and shrinking of gaseous nuclei.

14. The primary investigative technique of epidemiology is:
   a) computer modeling   b) library research
   c) reviewing data from patients   d) performing animal experiments

15. Which of the following is not a limitation of epidemiological studies?
   a) Medical charts are sometimes incomplete or inaccurate.
   b) Other factors, unrelated to the ultimate goal of the study, must be accounted for.
   c) When the bioeffect rate is small, numerous patients must be studied.
   d) Even when a bioeffect is identified, the investigator is still unsure of the specific reason for its occurrence.
   e) It does not provide greater understanding of the biologic significance of bioeffects.

16. True or False? The intensity limit established for diagnostic ultrasound is higher for focused sound beams than for broad, unfocused beams.

17. Which measure of intensity is related most closely to tissue heating?

    a) I_m b) SATA   c) SPPA   d) SPTA   e) SPTP

18. True or False? In vitro studies are performed exclusively on living animals.

19. True or False? Bioeffects identified through in vitro research are not considered real.

20. The next step following the identification of an in vitro bioeffect is:

    a) Propose in vivo research to evaluate the clinical significance of the bioeffect.
    b) Cease performing clinical studies that may induce bioeffects in vivo.
    c) Ignore the results: it's only an in vitro observation.
    d) Continue with in vitro research.
    e) Obtain an informed consent regarding this report from all patients.

21. True or False? Exposing an individual to ultrasound from a diagnostic imaging system is never appropriate when there is absolutely no clinical benefit to that individual.

22. Which component of an ultrasound system is most likely to expose a patient to danger?

    a) CRT   b) transducer   c) scan converter   d) pulser

23. Which is generally true of diagnostic ultrasound?

    a) Harmful bioeffects do not occur, and it is unnecessary to discuss them.
    b) Harmful bioeffects do not occur, but it is irresponsible to ignore the possibility that they may.
    c) Harmful bioeffects are often seen, but the benefit to patients still outweighs the risk.
    d) Only new technologies such as endovaginal or intravascular ultrasound need to be evaluated for bioeffects.

ANSWERS - BIOEFFECTS AND SAFETY

1 - c: One of the fundamental principles associated with the practice of medicine is balancing the benefit and risks to patients. A health care professional may proceed when the benefits surpass the risks.

2 - dosimetry

3 - False.

4 - False.

5 - True: This question and the preceding two define the scope of bioeffects. Bioeffects are indeed present; however, *harmful* bioeffects are not observed with typical diagnostic ultrasound.

6 - c: Evaluating patient charts is an element of the empirical study of bioeffects.

7 - Empirical approach. Bioeffects are being investigated in an "exposure-response" context.

8 - The two most important mechanisms associated with ultrasound-induced bioeffects are **thermal** and **cavitation**.

9 - d: Scientific conclusions are strongly justified when empirical and mechanistic investigations produce consistent results.

10 - True.

11 - False: Inertial, transient, and normal cavitation are the same. The correct answer is stable and normal.

12 - True.

13 - True: With stable cavitation, the microbubbles do not burst.

14 - c) reviewing data from patients.

15 - e: The greatest strength of epidemiological research is its dependence on clinical significance. If the bioeffect is observed in patients, then we are sure of its clinical relevance.

16 - True: Broad, unfocused sound beams cause greater tissue heating. The maximum intensity limit for these beams is therefore lower than that for focused beams.

17 - d: SPTA

18 - False: In vitro means literally "in glass."

19 - False: The AIUM conclusively states that in vitro bioeffects must be considered real; however, their clinical importance is unknown.

20 - a: Further in vivo research should be performed to understand the impact of in vitro-identified bioeffects on living subjects.

21 - False: It is appropriate to perform US studies for the sake of research and training even when no direct medical benefit is anticipated. Informed consent should be obtained from the subject.

22 - b: Generally, the transducer is the component with the greatest potential for exposing a patient to risk. The likelihood of injury is minuscule, but the potential exists.

23 - b: Diagnostic ultrasound has a remarkable safety record; however, we must always protect our patients from potential harm.

DECIBELS AND INTENSITY RATIOS

Decibel	Intensity ratio	Decibel	Intensity ratio
1	1.26	-1	0.79
2	1.58	-2	0.63
3	2.00	-3	0.50
5	3.16	-5	0.32
6	4.00	-6	0.25
10	10.0	-10	0.10
15	31.6	-15	0.032
20	100.	-20	0.01
25	316.	-25	0.0032
30	1000.	-30	0.001
40	10000.	-40	0.0001

The intensity ratio is the final intensity divided by the original intensity.

SINES AND COSINES OF CERTAIN ANGLES

Angle	Sine	Cosine
0	0.00	1.00
5	.087	.996
10	.173	.985
15	.258	.966
20	.342	.940
25	.422	.934
30	.500	.866
35	.573	.819
40	.642	.766
45	.707	.707
50	.766	.642
55	.819	.573
60	.866	.500
65	.934	.422
70	.940	.342
75	.966	.258
80	.985	.173
85	.996	.087
90	1.00	0.00

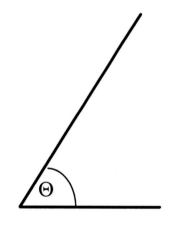

II. EXAM REVIEW

1. What term is used to describe the effects of an ultrasound wave on living tissues?
   a) toxic effects
   b) acoustic propagation properties
   c) biological effects
   d) transmission properties

2. As sound travels through a medium, what term describes the effects of the medium on the wave?
   a) toxic effects
   b) acoustic propagation properties
   c) bioeffects
   d) transmission properties

3. Select the sequence that appears in increasing order:
   a) mega, kilo, hecto, milli, giga
   b) nano, milli, micro, deci, deca, mega
   c) centi, deci, deca, hecto
   d) milli, hecto, centi, deci, nano, giga

4. The letters below represent the abbreviations for the prefixes of the metric system. Select the sequence that appears in decreasing order.
   a) m, k, M, g, da
   b) g, m, k, d, u
   c) g, k, di, m, u, n
   d) M, k, da, d, c

5. Match the following prefixes with their meanings.
   a) mega
   b) hecto
   c) milli
   d) kilo
   e) nano
   f) giga

   1) hundreds
   2) thousands
   3) thousandths
   4) millions
   5) billions
   6) billionths

6. Which of the following is not a measure of area?
   a) square cm
   b) meters squared
   c) cubic meters
   d) feet x feet

7. The perimeter of an anatomical structure is measured by a sonographer. Which of the following choices is a reasonable value for the measurement?
   a) $6 \, cm^2$
   b) 5 cc
   c) 15 mm
   d) 18 dB

8. What units are associated with the circumference of a circle?
   a) mm
   b) mm^2
   c) cm^3
   d) m^4

1.   **C.** The effects of ultrasound on the tissues are called biological effects or bioeffects. There have been no confirmed bioeffects on humans with acoustic intensities typical of those used in diagnostic imaging.

2.   **B.** Acoustic propagation properties describes the effect of the medium on the wave traveling through it. Acoustic means "sound." Propagation means "to travel."

3.   **C.** Centi means "one-hundredth", deci means "one-tenth", deca means "ten" and hecto means "hundred." Thus, this is a sequence of increasing numbers.

4.   **D.** M stands for mega and means "millions." k stands for kilo and means "thousands." da stands for deca and means "tens." d stands for deci and means "one-tenth." c stands for centi and means "one-hundredth." Thus, this is sequence of decreasing numbers.

5.   The correct matches are as follows: **a & 4, b & 1, c & 3, d & 2, e & 6, and f & 5.** You should be familiar with the abbreviations, prefixes, and meanings for all of the terms associated with the metric system.

6   C. Cubic meters are not used to measure area. An area is measured in units of distance squared. For example, a rectangle's area is its length multiplied by its width and can be reported in square feet, in^2, or square miles.

7.   C. The term perimeter describes the length of the outer boundary of a structure. For example, a square with an edge that is 5 inches long will have a perimeter measuring 20 inches (four sides, each with a length of 5 inches). Thus, the answer is C. Units of measurement in A, B, and D are not applicable to length.

8.   **A.** The circumference of a circle is the length of the circle's outer boundary and thus is measured in units of length. Selections B and C are units of area and volume. Choice D, with units of length to the fourth power, has no meaning in geometry.

9. The volume of a cystic structure is estimated from sonographic data. Which of the following is an acceptable measurement of this volume?
   - a) 6
   - b) 6 cm
   - c) 6 cm^2
   - d) 6 cm^3

10. The speed of red blood cells traveling through a blood vessel is 750 cm/sec. You are asked to convert this measurement of speed to miles per hour. What information would you need?
    - a) the number of seconds in a minute and the number of blood cells in the vessel
    - b) the number of miles in a meter
    - c) the number of seconds per hour and the number of miles in a centimeter
    - d) the direction of red blood cell motion and the Doppler shift frequency

11. How many nanoseconds are in 7 seconds?
    - a) 7,000,000
    - b) 7,000,000,000
    - c) 7 million
    - d) 0.00000007

12. How are scores on the ultrasound board exams related to the time a sonographer devotes to studying for them?
    - a) inversely
    - b) conversely
    - c) directly
    - d) unrelated

13. Sound can be characterized as _____.
    - a) energy flowing through a vacuum
    - b) a variable
    - c) cyclical oscillations in certain variables
    - d) a principle of acoustics

14. Which of the following is true of all waves?
    - a) they travel through a medium
    - b) all carry energy from one site to another
    - c) their amplitudes do not change
    - d) they travel in a straight line

9.   **D.** The units of volume are length cubed, such as ft^3 or cubic centimeters. This may also be expressed as length to the third power. The only option with these units is D. A has no units, while B and C have units of length and area.

10. **C.** To convert one unit to another requires a factor that relates the two terms. For example, to change a measurement of length from feet into inches, we must know how many inches there are in a foot. In this case, we are asked to convert cm per sec into miles per hour. Therefore, we require two pieces of information: the relationship between seconds and hours (relating one unit of time to another) and the relationship between miles and centimeters (relating one unit of distance to another). *When units are changed, the actual amount does not change. For* example, ten dimes gives us the same purchasing power as four quarters. The units may be different, but the "total picture" remains the same.

11. **B.** Nano means billionth. There are one billion billionths in one second. Therefore, there are seven billion billionths in seven seconds.

12. **C.** Two variables are related when changes in one variable result in changes in the other. When the changes follow in the same direction (both increase or both decrease), then the variables are directly related. Because, scores increase as studying time increases, these variables are directly related.

13. **C.** Sound is a wave. A wave is the rhythmical variation throughout time.

14. **B.** Waves carry energy from one place to another. A is incorrect because some types of waves, such as light waves, can travel through a vacuum. C is incorrect because many waves get weaker as they travel. Certain waves do not travel in a straight line; thus D is also incorrect.

15. A longitudinal wave is propagating from the East toward the West at a speed of 2 miles per hour. What is the direction of motion of the particles within the wave?
    a) from the East to the West only
    b) alternately from East to West and then from West to East
    c) from North to South only
    d) alternately from South to North and then from North to South

16. A particle within a transverse wave is traveling vertically. What is the direction of the wave's propagation?
    a) horizontal   b) vertical
    c) diagonal; both horizontal and vertical   d) it cannot be determined

17. Which of the following types of waves do not require a medium in order to propagate? (More than one answer may be correct.)
    a) light   b) heat
    c) sound   d) television

18. Which of the following describes the characteristics of a sound wave?
    a) longitudinal, non-mechanical   b) mechanical, transverse
    c) transverse, acoustic   d) mechanical, longitudinal

Each of the following terms is an acoustic variable. True or False?

19. frequency

20. pressure

21. propagation speed

22. wavelength

23. temperature

24. intensity

25. motion of particles in the wave

26. density

15. **B.** A longitudinal wave is defined as a wave whose particles vibrate back and forth in the same direction that the wave is propagating. Therefore, since the wave is traveling from the East to the West, the particles in this wave will vibrate repeatedly from East to West and then from West to East.

16. **A.** The particles within a transverse wave travel in a direction that is perpendicular to the direction in which the wave itself is traveling. If a transverse wave is traveling vertically, the particles in the wave are traveling horizontally. A water wave is a primarily a transverse wave. The wave propagates sideways along the surface of the water, whereas a ball floating on top of the water moves up and down as the wave passes.

17. **A, B, and D.** Sound cannot travel through a vacuum; it requires a medium in order to propagate. Other waveforms such as light, heat and TV waves *are* capable of traveling through a vacuum.

18. **D.** Sound is both a mechanical wave and a longitudinal wave. A mechanical wave, such as sound, actually imparts energy to the molecules of the medium through which it travels. The molecules of the medium vibrate, striking their neighbors, which in turn vibrate. This chain reaction results in the acoustic energy traveling through the medium.

19. **False.** Acoustic variables are measured quantities whose values change as a sound wave propagates. These quantities are pressure, temperature, density, and the motion of particles in the wave.

20. **True.**

21. **False.**

22. **False.**

23. **True.**

24. **False.**

25. **True.**

26. **True.**

27. Which of the following units can be used to report the pressure measurement of an acoustic wave? (More than one answer may be correct)
    a) atmospheres (atm)   b) Pascals (Pa)
    c) millimeters of mercury (mm Hg)   d) pounds/sq. inch (lb/in^2)

28. A force is applied to a surface. If the force is tripled and, at the same time, the surface area over which the force is applied is also tripled, what is the new pressure?
    a) three times larger than the original   b) one third of the original
    c) six times more than the original   d) unchanged

29. A sound wave propagating in air can be considered to be a rhythmical compression and rarefaction of air molecules as depicted below. Where is the location of highest density?

Temperature is an acoustic variable. Are the next three statements true or false?

30. Temperature is a measurement of the energy in a wave.
31. Heat flows from areas of low temperature to areas of high temperature.
32. The units of "degrees" are acceptable as an acoustic variable.

33 The sketch below is an example of a(n) _____ wave.

    a) sound   b) electromechanical
    c) transverse   d) longitudinal

34. Which of the following units are appropriate to describe the period of an acoustic wave? (More than one answer may be correct.)
    a) minutes   b) microseconds
    c) meters   d) mm/us   e) deciliters

27. **A, B, C, and D.** All of these terms are appropriate to represent pressure (just as weight can be reported with units of pounds, ounces, tons, or grams).

28. **D.** Pressure is defined as an amount of force divided by the area to which it is applied. If the applied force is tripled and, at the same time, the area over which it is applied is tripled, then the pressure remains unchanged.

29. **D.** The highest density is located where a volume contains the largest number of molecules. This occurs in the figure at letter D.

30. **True.** Temperature is a measurement of the total energy of an object.

31. **False.** Heat flows from areas of higher temperature to areas of lower temperature. That is why we feel warmth when sitting near a fire. The heat moves from the fire toward us.

32. **True.** Temperature is an acoustic variable. Since temperature is measured in degrees, these units are appropriate as an acoustic variable.

33. **C.** The sketch in the question is a transverse wave. The particles in the wave move up and down, while the wave itself moves sideways.

34. **A and B.** The period of a wave is defined as the time that elapses as a wave oscillates through a single cycle. The units of period must be a measure of time, such as minutes or seconds. Choices A and B are units of time. The incorrect selections C, D, and E are units of distance, speed, and volume, respectively.

Match the four acoustic variables with their correct units. (More than one
   answer may be correct.)

35. Pressure   1. lb/in^3
36. Temperature   2. Fahrenheit degrees
37. Density   3. miles
38. Particle motion   4. Pa
                                      5. cm
                                      6. centigrade degrees
                                      7. lb/in^2
                                      8. kg/m^3

39. _____ is the reciprocal of period.
    a) inverse period   b) pulse repetition period
    c) frequency   d) propagation period

40. What is the range of periods commonly found in waves produced by
    ultrasound systems?
    a) 0.001 to 1 sec   b) 0.1 to 1 μsec
    c) 0.1 to 1 msec   d) 10 to 100 nsec

41. With standard ultrasonic imaging, what happens to the period of a wave as
    it propagates?
    a) increases   b) decreases
    c) remains the same

42. True or False? If the periods of two waves are the same, then the
    frequencies of the waves must also be the same.

43. True or False? The sonographer has the ability to alter the period of an
    ultrasound wave that is produced by a transducer typically used in
    diagnostic imaging.

35. **4 and 7.** Units of pressure include Pascals (Pa) and pounds per square inch (lb/in^2).

36. **2 and 6.** Units of temperature include both Fahrenheit and centigrade degrees.

37. **1 and 8.** Density is the concentration of mass per unit of volume. Mass is measured in pounds (lb) or kilograms (kg). Volume has units of length cubed, such as cubic inches (in^3) or cubic meters (m^3). Therefore, the correct answers are 1 and 8, lb/in^3 and kg/m^3.

38. **3 and 5.** Particle motion is measured in units of distance, such as miles or cm. These are the only choices that represent measures of distance.

39. **C.** Frequency is the reciprocal of period. Mathematical reciprocals are related in the following manner: first, as one increases, the other decreases; second, when the two numbers are multiplied together, the result is unity. For example, a wave with a period of one-hundredth of a second has a frequency of 100 per second or 100 Hz..

40. **B.** Ultrasonic imaging waves have a period in the range of 0.1 to 1.0 µsec. The period is the time that passes as an acoustic variable goes through a single cycle. Period is the reciprocal of the frequency. A wave with a frequency of 1 MHz has a period of 1 µsec. A wave with a frequency of 10 MHz has a period of 0.1 µsec.

41. **C.** Certain parameters of a wave change as the wave travels through the body. However, the period, as well as the frequency, of a wave typically remains constant as a sound wave propagates.

42. **True.** Frequency and period are reciprocals. When the periods of two waves are identical, the frequencies of the waves must also be identical.

43. **False.** In diagnostic imaging, the sonographer cannot adjust the period (or frequency) of a wave that is produced by a particular transducer. If the sound beam's frequency and period are not compatible with a particular type of imaging, the sonographer must select a new transducer with a different frequency.

44. What determines the period of an ultrasound wave?
    a) the transducer   b) the medium through which the sound travels
    c) both a and b   d) neither a nor b

45. What term describes the number of cycles completed by an acoustic variable in a second?
    a) period   b) frequency
    c) pulse repetition period d) variable rate

46. Which of the following cannot be considered a unit of frequency?
    a) per day   b) cycles/sec   c) Hz
    d) Hertz   e) cycles

47. What is the range of frequencies emitted by transducers used in ultrasonic imaging?
    a) 1 to 3 MHz   b) 1 to 1,000 kHz
    c) -10,000 to +10,000 Hz   d) 2,000,000 to 10,000,000 Hz

48. True or False? In current diagnostic imaging instruments, a sonographer can change the frequency of sound emitted by the transducer's ultrasonic crystal.

49. What establishes the frequency of an ultrasound wave?
    a) the transducer   b) the medium through which the sound travels
    c) both a and b   d) neither choice a nor b

50. True or False? With standard ultrasound pulses, the frequency of the ultrasound changes significantly as the wave propagates through the body.

44. **A.** The sound source (the transducer) that produces an acoustic signal determines the period of a wave. The wave's period is independent of the medium through which the sound travels, and will not change as the wave progresses from one medium to another.

45. **B.** Frequency may also be described as the rate of recurrence or the number of events that regularly occur in one second.

46. **E.** The term cycles informs us of the number of events, but does not inform us of the duration of time required for those events to occur. Therefore, choice E is incomplete, and is not a unit frequency. All of the other choices reveal that a number of events took place in a specific time span.

47. **D.** Frequencies commonly used in diagnostic imaging range from approximately 2 to 10 megahertz or 2 to 10 million cycles per second.

48. **False.** Imaging transducers used most commonly with today's ultrasound systems emit acoustic energy at one primary frequency. In order to alter the primary frequency used in an exam, the sonographer must select a different transducer. Some systems appear to allow the sonographer to change the frequency of the transducer, but this is not the case. These systems have multiple ultrasound crystals located in the transducer housing. When the sonographer changes the frequency, he is actually selecting a different crystal within the assembly.

49. **A.** When created by a transducer, an ultrasound pulse has a certain and specific frequency. The frequency of the pulse is not determined by the medium through which the sound travels. Only the sound source (the transducer) establishes the wave's frequency.

50. **False.** In diagnostic imaging, the frequency of the sound wave generally remains constant and does not routinely change as the sound propagates through the body. Slight changes in frequency occur when sound strikes moving structures. This forms the basis for Doppler ultrasonography.

50. Ultrasound is defined as a sound with a frequency of _____.
    a) greater than 20,000 kHz   b) less than 1 kHz
    c) greater than 10 MHz   d) greater than 0.02 MHz

51. Infrasound is defined as a sound with a frequency of _____.
    a) greater than 20,000 kHz   b) less than 20 Hz
    c) greater than 10 MHz   d) less than 0.02 MHz

52. True or False? Waves in the ultrasound range generally behave the same as sound waves that are audible.

53. What is characteristic of acoustic waves with frequencies exceeding 20,000 Hz when compared with waves having frequencies of less than 20,000 Hz?
    a) they travel more effectively in soft tissue
    b) they travel more rapidly
    c) they attenuate less when traveling in soft tissue
    d) humans cannot hear them

54. What is characteristic of acoustic waves with frequencies less than 20 Hz when compared with waves having frequencies of more than 20 Hz?
    a) they travel less effectively in soft tissue
    b) they travel more rapidly
    c) they attenuate more when traveling in soft tissue
    d) humans cannot hear them

55. Let us compare two acoustic waves, A and B. The frequency of wave A is one-third the frequency of wave B. How does the period of wave A compare with the period of wave B?
    a) A is one-third as long as wave B   b) A is the same as wave B
    c) A is three times longer than wave B   d) it cannot be determined

56. What determines the initial amplitude of an ultrasound wave?
    a) the transducer   b) the medium through which the sound travels
    c) both a and b   d) neither choice a nor b

57. True or False? With standard diagnostic imaging instrumentation, the sonographer has the ability to vary the amplitude of a sound wave produced by the transducer.

50. **D.** Ultrasound is defined as an acoustic wave with a frequency so *high* that it is not audible to humans. Ultrasound is an inaudible wave with a frequency of at least 20,000 Hertz or 0.02 MHz.

51. **B.** Infrasound is defined as an acoustic wave with a frequency so *low* that it is not audible to humans. Infrasound is an inaudible wave with a frequency less than 20 Hz.

52. **True.** As stated above, the primary difference between audible and ultrasonic waves is that humans can hear audible waves. A wave's behavior or adherence to physical laws and principles is the same, regardless of whether or not it can be heard by humans.

53. **D.** Ultrasonic waves have frequencies exceeding 20 kHz and are inaudible to humans. They travel at the same speed as waves with lower frequencies and attenuate at a greater rate than waves at lower frequencies.

54. **D.** Infrasonic waves have frequencies less than 20 Hz and are inaudible to humans. They travel at the same speed as waves with higher frequencies and attenuate at a lesser rate than waves at higher frequencies.

55. **C.** Frequency and period are reciprocals. Therefore, if the frequency of one wave is one-third as large as another, then the period of that wave will be three times longer than the other.

56. **A.** The initial strength or amplitude of a sound wave is determined by the vibration of the piezoelectric crystal in the transducer. The greater the vibration of the crystal, the larger the amplitude of the ultrasound wave.

57. **True.** The sonographer can vary or adjust the strength of the ultrasound signal created by a transducer. This is usually achieved by adjusting a control on the system's console. When a sonographer increases the output power, the electrical voltage transmitted to the transducer is increased. This produces a more violent vibration of the piezoelectric crystal within the transducer and, in turn, a stronger ultrasound wave.

58. As an ultrasound wave travels through the body, its amplitude
    usually _____ .
    a) decreases   b) increases
    c) remains the same   d) it cannot be determined

59. Which of the following are acceptable units for the amplitude of an
    acoustic wave? (More than one answer may be correct.)
    a) cm   b) atmospheres
    c) degrees   d) watts

60. What is the amplitude of the wave depicted below?

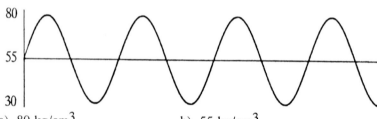

    a) 80 kg/cm^3 b) 55 kg/cm^3
    c) 30 kg/cm^3 d) 25 kg/cm^3

61. The maximum value of the density of an acoustic wave is 60 lb/in^2
    whereas the minimum density is 20 lb/in^2. What is the wave's amplitude?
    a) 20 lb/in^2 b) 40 lb/in^2
    c) 60 lb/in^2 d) none of the above

62. The power of an ultrasound wave can be reported with which of the
    following units? (More than one answer may be correct.)
    a) watts/square centimeter   b) dB
    c) watts   d) kg/cm^2

63. Two waves arrive at the same location and interfere. The resultant sound
    wave is smaller than either of the two original waves. What is this called?
    a) constructive interference   b) angular interaction
    c) destructive interference   c) in-phase waves

64. A pair of waves are in phase. If these waves interfere with each other,
    what will occur?
    a) reflection   b) constructive interference
    c) refraction   d) destructive interference

58. **A.** As a sound wave travels in the body, its strength or amplitude diminishes. This process is called attenuation. We experience attenuation when we walk away from a person who is speaking. The further away we are from a speaker, the weaker their voice becomes, and the more difficult it is for us to hear them.

59. **A, B, and C .** The acoustic variables are pressure, density, particle motion and temperature. The units of acoustic wave amplitude are: cm for particle motion, atmospheres for pressure, and degrees for temperature. D is incorrect because watts are units of power, which is not an acoustic variable.

60. **D.** The amplitude of a wave is calculated by subtracting the average value of the acoustic variable from its maximum value. If the maximum value is 80 kg/cm^3 and the average value is 55 kg/cm^3, the amplitude is determined as follows: 80 - 55 = 25 kg/cm^3.

61. **A.** The amplitude of a wave may be calculated by subtracting the minimum value of the acoustic variable from its maximum and then dividing that number in half. In this case, the maximum minus the minimum as follows:
60 - 20 = 40 lb/in^2. Half of 40 lb/in^2 is 20 lb/in^2.

62. **C.** The units of power are watts. This same measure of power is used for light bulbs, stereo systems, and curling irons.

63. **C.** Destructive interference results when a pair of out of phase waves interfere with each other. The sum of the two out of phase waves has a smaller amplitude than either of the original waves.

64. **B.** Waves that are in phase constructively interfere with each other. The single wave that results from the combination of the two in phase waves will always have a higher amplitude than either of the original waves.

65. Typically, as an ultrasound wave travels through soft tissue, the power of the wave _____.
   a) decreases   b) increases
   c) remains the same

66. True or False? A sonographer can routinely change the power of a wave emitted by a transducer used in diagnostic ultrasonic imaging.

67. Mathematically, when a number is squared, it means that the number is multiplied by _____.
   a) 2   b) 0.5
   c) itself   d) 1.5

68. The final amplitude of an acoustic wave is reduced to one-half of its original value. The final power is _____ the original power.
   a) the same as   b) one-half of
   c) double   d) none of the above

69. The amplitude of an acoustic wave decreases from 27 degrees to 9 degrees. If the initial power in the wave was 27 watts, what is the wave's final power?
   a) 3 watts   b) 9 watts
   c) 1 watt   d) none of the above

70. The intensity of an ultrasound beam is defined as the _____ in a beam _____ by the _____ of the beam.
   a) power, multiplied, diameter   b) amplitude, divided, area
   c) power, divided, area   d) amplitude, multiplied, circumference

71. As sound travels in the body, what typically happens to the intensity of the wave?
   a) increases   b) decreases
   c) remains the same

72. What are the units of intensity?
   a) watts   b) watts/cm
   c) watts/cm^2 d) dB

65. **A.** As a sound wave travels through the body, its power diminishes. This process is called attenuation. Amplitude and power are both measures of the strength of an acoustic wave and tend to decrease as sound travels.

66. **True.** A sonographer can alter the power of an ultrasound wave by adjusting a control on the ultrasound system. Power and amplitude are related; if the amplitude is increased, then so is the power. When the amplitude decreases, the power also decreases.

67. **C.** To square a number means to multiply the number by itself. Hence, the term "five squared" means 5 times 5, or 25. Ten squared is 10 times 10, or 100.

68. **D.** Changes in the power of a wave are proportional to changes in the wave's amplitude squared. One-half squared equals one-quarter (one-half times one-half equals one-quarter). When one-half of the wave's original amplitude remains, then only one-fourth of the original power remains.

69. **A.** Alterations in a wave's power are proportional to changes in its amplitude squared. The power is reduced to one-third of its previous value (from 27 to 9 degrees). One-third squared equals one-ninth. Thus, only one-ninth of the wave's original power remains. The initial power in the wave was 27 watts; one-ninth of that is 27/9, or 3 watts.

70. **C.** The intensity of an ultrasound beam is the concentration of the power within the beam area. It is calculated by dividing the power in a beam by its area. The units of intensity are watts per centimeter squared.

71. **B.** The intensity of a sound beam decreases as it travels through the body because of attenuation. Amplitude, power, and intensity are three different ways to measure the strength of an ultrasound beam. They all decrease as sound propagates.

72. **C.** The intensity of a beam is the power in the beam divided by its cross-sectional area. Power has units of watts, and beam area has units of square cm. Intensity therefore has units of watts/cm^2.

73. When the power in an acoustic beam is doubled and the cross-sectional area of the beam is halved, then the intensity of the beam is _____ .
    a) doubled   b) halved
    c) quartered   d) four times larger

74. If the power in an ultrasound beam is unchanged while, at the same time, the beam area doubles, then the beam's intensity _____ .
    a) doubles   b) is halved
    c) is quartered   d) remains the same

75. What happens to an acoustic beam's intensity when the power in the beam increases by 25% while the cross-sectional area of the beam remains the same?
    a) it increases by 25%   b) it increases by 75%
    c) it increases by 50%   d) it decreases by 25%

76. True or False? With diagnostic ultrasonic imaging instruments, the operator can alter the intensity of an ultrasound beam produced by a transducer.

77. What determines the initial intensity of an ultrasound beam?
    a) the source of the sound wave
    b) the medium through which the sound travels
    c) both a and b
    d) neither choice a nor b

73. **D.** Intensity is equal to power divided by beam area. In this case, the power is doubled, and the area is halved. The intensity rises to four times its original value. For example, if the original power was 4 watts and the initial beam area was 4 cm^2, then the starting intensity was 4 watts divided by 4 cm^2, or 1 watt/cm^2. Now, the power is doubled, from 4 to 8 watts, and the beam area is halved, from 4 to 2 cm^2. The new intensity is 8 watts divided by 2 cm^2, or 4 watts/cm^2. The initial intensity is 1, and the final intensity is 4 watts/cm^2; therefore, the intensity has increased fourfold.

74. **B.** Intensity is equal to power divided by beam area. In this case, the power is unchanged while the beam area is doubled. Therefore, the beam's intensity is halved. For example, if the original power was 2 watts and the initial beam area was 2 cm^2, then the starting intensity was 2 watts divided by 2 cm^2, or 1 watt/cm^2. Now, beam area is doubled, from 2 to 4 cm^2. The new intensity is 2 watts divided by 4 cm^2 or 0.5 watts/cm^2. The initial intensity is 1, and the final intensity is 0.5 watts/cm^2; therefore, the intensity has been halved.

75. **A.** The intensity is equal to the power divided by the beam area. If the power increases by 25% while the beam's area is unchanged, then the intensity is also increased by 25%.

76. **True.** Although the operator cannot change all of the characteristics of an ultrasound beam created by an individual transducer, he or she can change the power radiated from it. As the operator increases the output power of a transducer, its intensity increases.

77. **A.** The source of the acoustic wave determines its initial intensity (as well as the wave's amplitude and power). At its point of origin, the strength of an acoustic wave is not related to the medium that the sound is about to enter.

78. What determines the intensity of an ultrasound beam after it has traveled through the body?
    a) the wave's source   b) the medium through which the sound travels
    c) both a and b   d) neither choice a nor b

79. What happens to the intensity of an ultrasound beam when the beam's cross-sectional area remains unchanged while the amplitude of the wave triples?
    a) it triples   b) it increases ninefold
    c) it remains the same   d) none of the above

80. If the power of the beam is tripled while the cross-sectional area of the beam remains the same, the intensity _____.
    a) triples   b) increases ninefold
    c) remains the same   d) none of the above

81. The amplitude of an acoustic wave is increased. Which of the following will most likely remain unchanged? (More than one answer may be correct.)
    a) power   b) frequency
    c) period   d) intensity

82. The wavelength of a cycle in an ultrasound wave can be reported with which units?
    a) units of time (sec, min, etc.) b) units of distance (feet, etc.)
    c) units of area (m^2, etc.)   d) mm only

83. The wavelength of an ultrasonic wave is determined by _____.
    a) the sound source   b) the medium through which the wave travels
    c) both a and b   d) neither a nor b

84. What is the best estimate of the distance that sound can travel in soft tissue in one second?
    a) one yard   b) one hundred yards
    c) one mile   d) ten miles

78. **C.** The combination of the source and the medium ultimately determine the residual intensity of an ultrasound beam following a passage through the body. As an acoustic wave propagates, attenuation occurs. The initial intensity of a sound beam is established by the source of the sound, the transducer. The frequency, which is also determined by the transducer, impacts how much attenuation occurs. In addition, the characteristics of the medium help to determine attenuation. For example, bone and lung have a greater attenuation rate than that of soft tissue. In contrast, water has a lower attenuation rate than soft tissue.

79. **B.** A change in the intensity of a wave is proportional to the change in the amplitude of the wave squared. When the amplitude of a wave is tripled, the intensity is increased ninefold (3 x 3 = 9).

80. **A.** When the cross-sectional area of a beam remains unchanged (which is true in this question), the change in the intensity of a beam is proportional to the change in the power. The question states that the power is tripled. If the power is tripled, then the intensity is also tripled.

81. **B and C.** Amplitude relates to the strength of a wave. The frequency of a wave describes the number of times the wave cycles in one second. A wave's period is the time that passes as an acoustic variable oscillates though one complete cycle. Therefore, frequency and period are not related to amplitude and remain unaltered when the amplitude changes.

82. **B.** Wavelength is the *distance* from the beginning of a cycle to the end of that cycle. It is a length, and therefore has units of distance. In soft tissue and with frequencies typical of diagnostic imaging, wavelengths range from 0.15 to 0.75 mm. Although it may be impractical to record wavelengths in miles or meters, it can be done.

83. **C.** Wavelength is determined by both the sound source and the medium through which the sound travels. Wavelength is determined, in part, by the wave's frequency (which is established by the sound source). Wavelength is also affected by the propagation speed of the sound wave, which is established by the medium. In a particular medium, the higher the wave's frequency, the shorter the wavelength. With a particular frequency, the faster the propagation speed, the longer the wavelength.

84. **C.** Sound travels at a speed of 1,540 meters per second in soft tissue. This is approximately one mile per second.

An ultrasound pulse propagates from soft tissue through a cyst. The speed of sound through the cyst is 1,575 meters per sec. Are the following four statements true or false?

85. The frequency of the wave increases as it travels through the cyst.
86. The period of the sound wave decreases as it travels through the cyst.
87. The wavelength increases while the wave travels through the cyst.
88. The power in the wave increases as it travels through the cyst.

89. The speed of sound traveling through bone is _____ that in soft tissue.
    a) higher than   b) lower than
    c) equal to   d) it cannot be determined

90. Compared with soft tissue, the speed of an acoustic wave through lung tissue is _____ .
    a) faster   b) slower
    c) equal   d) it cannot be determined

91. What is the speed of sound in air?
    a) 1,540 m/s   b) 330 m/s
    c) 100 m/s   d) 3,010 m/s

92. True or False? Soft tissue is an imaginary construct that actually does not exist.

93. What is the propagation speed of a 5 megahertz sound wave in soft tissue?
    a) 1,450 meters per sec   b) 1,540 km/sec
    c) 1.54 m/s   d) 1.54 mm/μs

94. What two properties establish the propagation speed of sound in a given medium?
    a) elasticity and stiffness   b) stiffness and impedance
    c) conductance and density   d) density and stiffness

85. **False.** The frequency of a wave is determined by the sound source only, and is unaffected by the medium or media through which it travels.

86. **False.** The period of a wave is determined by the sound source only and is unaffected by the medium or media through which it travels. (Also, period and frequency are reciprocals. Since the frequency remained constant, the period could not change.)

87. **True.** Wavelength is dependent on both the sound source and the medium through which the sound travels. The speed of sound in soft tissue is 1,540 m/sec; and through the cyst it is 1,575 m/sec. As a wave travels in a faster medium, its wavelength increases.

88. **False.** Typically, the power of a beam diminishes as it travels. This is a result of attenuation. If anything, the wave's power diminishes as it propagates through the cyst.

89. **A.** The propagation speed of sound in bone is higher than that in soft tissue. Sound travels at a speed of 3 to 5 km/sec in bone.

90. **B.** The speed of sound in lung tissue is slower than that in soft tissue. Sound travels at speeds in the range of 0.5 to 1.0 km/sec in lung.

91. **B.** The speed of sound in air is 330 m/s, substantially lower than the speed of sound in soft tissue.

92. **True.** Soft tissue is an imaginary structure with characteristics that represent an "average" of body tissues, including the muscle, blood, kidney, and spleen. Soft tissue is used as an approximation of the tissues through which ultrasound beam will pass during a diagnostic exam.

93. **D.** The speed of *any* sound wave moving through soft tissue, regardless of frequency, is 1,540 m/sec, or 1.54 km/sec, or 1.54 mm/μsec.

94. **D.** The propagation speed of an ultrasound wave is determined by the characteristics of the medium through which the sound travels. The properties of the medium that affect its propagation speed are density and stiffness.

95. Compressibility is a characteristic of a medium. Which two terms describe the same property as compressibility?
   a) density and stiffness   b) density and conductance
   c) stiffness and elasticity   d) elastance and impedance

96. The characteristics of four media are described below. Which of the media has the lowest propagation speed?
   a) high density and high elasticity b) low density and high stiffness
   c) low stiffness and low density   d) low compressibility & low stiffness

97. The speed at which a wave travels through a medium is determined by:
   a) the sound wave's properties only   b) the medium's properties only
   c) properties of both wave & medium   d) none of the above

98. When the elasticity of a medium is high, the _____ is high.
   a) stiffness   b) propagation speed
   c) compressibility   d) reflectivity

99. Two sound waves, one with a frequency of 5 megahertz and the other with a frequency of 3 megahertz, travel to a depth of 8 cm in the same medium and then reflect back to the surface of the body. Which acoustic wave arrives first at the surface of the body?
   a) the 5 MHz wave   b) the 3 MHz wave
   c) neither   d) it cannot be determined

100. Propagation speed can be correctly recorded with which of the following units? (More than one answer may be correct.)
   a) miles per hour   b) mm/μsec
   c) km/sec   d) inches per year

101. If sound travels at exactly 1,540 m/sec in a particular medium, then the medium _____.
   a) must be soft tissue   b) may be soft tissue
   c) cannot be soft tissue

95. **C.** The compressibility of a medium describes the ability to reduce its volume when a force is applied to it. For example, a marshmallow is compressible because when it is squeezed, it gets smaller. Elasticity has the same meaning as compressibility. Stiffness describes this same characteristic; however, stiffness is the opposite of compressible.

96. **A.** Speed is determined by density and stiffness of a medium (the opposite of stiffness is elasticity). When a medium has high elasticity, its speed is low. A medium with a high density also has a low speed. Therefore, the medium described in choice A has the lowest propagation speed.

97. **B.** Speed is determined by the characteristics of the medium only. The characteristics of the wave do not affect its speed. Therefore, all sound waves of any frequency, period, intensity, and power travel at the same speed in a particular medium.

98. **C.** Both elasticity and compressibility describe the ability of a medium to reduce its volume when affected by force. If the elasticity of a medium is high, so too is the compressibility.

99. **C.** They both travel at the same speed and reach the surface of the body at exactly the same time. *All sound waves, regardless of their features, travel at the same speed in a specific medium.* The different frequencies of these waves are irrelevant.

100. **A, B, C, and D.** Speed is recorded as a distance per unit of time, such as miles per hour, or feet/sec. Any relationship of distance divided by time is an acceptable answer.

101. **B.** Sound waves travel at 1,540 m/sec in soft tissue. Therefore, this medium could be soft tissue. However, there are other media with the same propagation speeds as those in soft tissue. So, the medium in this case may be soft tissue, or it may be something else.

102. If a sound wave doesn't travel at 1,540 m/sec in a medium, then the
     medium _____.
     a) must be soft tissue   b) may be soft tissue
     c) cannot be soft tissue

103. How long does it take for sound to make a round trip of 1 cm in soft
     tissue?
     a) 13 μsec   b) 150 msec
     c) 15 μsec   d) 2 seconds

104. The propagation speeds of ultrasound waves in muscle, liver, kidney, and
     blood are _____.
     a) exactly the same   b) very similar to each other
     c) vastly different

Two sound pulses travel through the same medium. One wave has a frequency
     of 2 MHz and the other has a frequency of 10 MHz. Considering these
     frequencies, answer the following five questions.

105. Which pulse has a longer wavelength?
     a) the 10 MHz pulse   b) the 2 MHz pulse
     c) neither pulse   d) it cannot be determined

106. Which pulse has a lower propagation speed?
     a) the 10 MHz pulse   b) the 2 MHz pulse
     c) neither pulse   d) it cannot be determined

107. Which pulse has a greater period?
     a) the 10 MHz pulse   b) the 2 MHz pulse
     c) neither pulse   d) it cannot be determined

108. Which pulse has a smaller power?
     a) the 10 MHz pulse   b) the 2 MHz pulse
     c) neither pulse   d) it cannot be determined

109. Which pulse has a longer spatial pulse length?
     a) the 10 MHz wave   b) the 2 MHz wave
     c) neither wave   d) it cannot be determined

102. **C.** Sound waves travel 1,540 m/sec in soft tissue. Since the propagation speed of sound in this medium is not 1,540 m/sec, the medium cannot be soft tissue.

103. **A.** In soft tissue, sound can travel to and return from a depth of 1 cm (a total distance of 2 cm) in 13 microseconds (13 millionths of a second).

104. **B.** The difference in the speed of sound between these media is less than 5%. The characteristics of muscle, kidney, liver, and blood that determine the wave's speed in the media are quite similar.

105. **B.** In any specific medium, the wave with the lower frequency has the longer wavelength. The 2 MHz wave has a longer wavelength than the 10 MHz wave.

106. **C.** The propagation speeds of all sound waves are identical while traveling in a specific medium. Therefore, neither wave travels at a lower speed; their speeds are identical.

107. **B.** Frequency and period are reciprocals. The wave with a greater frequency has a lower period. The wave with the lower frequency has the greater period. Therefore, the 2 MHz wave has the greater period.

108. **D.** The power of a wave is not related to its frequency. The power relates to the strength of the wave, and in this case, no information is provided about the power. Therefore, we cannot answer this question based on the information provided.

109. **D.** The spatial pulse length is equal to the wavelength multiplied by the number of cycles in the pulse. We know that 2 MHz sound has a longer wavelength than 10 MHz ultrasound. However, the number of cycles in each pulse is unknown. Therefore, the information provided is insufficient to answer the question.

Three sound waves with an identical frequency of 3 MHz have powers of 2 mW, 5 mW, and 15 mW. They all travel though three media -- wood, brick and fat, with identical thicknesses of 5 cm. Are the following three statements true or false?

110. The waves travel through all three media at the same speed since they have identical frequencies.

111. The sound waves travel through all three media at different speeds because the waves have different powers.

112. The waves travel through all three media at different speeds because the media are different.

113. The pulse repetition frequency (PRF) of ultrasound produced by a transducer typical of diagnostic imaging systems _____.
    a) can be changed by the sonographer
    b) depends upon the medium through which the sound travels
    c) remains unchanged as long as the same ultrasound system is used
    d) has nothing to do with ultrasonic imaging

114. The pulse repetition frequency (PRF) is used to describe certain characteristics of ultrasonic imaging systems. What are its units?
    a) seconds   b) 1/second
    c) mm/us   d) seconds^{-2}

115. In diagnostic imaging, what establishes the pulse repetition frequency of a pulsed wave?
    a) the sound source   b) the medium through which the pulse travels
    c) both a and b   d) neither a nor b

116. When a sonographer increases the maximum imaging depth during an exam, what happens to the PRF?
    a) increases   b) frequency decreases
    c) remains unchanged

110. **False**. The determinant of the propagation speed in a medium is the medium itself. Therefore, the fact that these three waves have the same frequency is irrelevant. The waves are traveling in media with vastly different densities and stiffnesses and as a result will travel at different speeds through the wood, brick, and fat.

111. **False**. The dissimilar powers of the waves do not affect their speeds. The waves travel at different speeds due to the dissimilarities in the densities and stiffnesses of the media through which they travel.

112. **True**. This is correct! Finally we can attribute variations in propagation speeds to the differences in the media. Again, the characteristics of the media alone determine the speed of sound traveling through them.

113. **A**. The pulse repetition frequency is related to the maximum imaging depth achieved during an exam. As the maximum imaging depth is increased by the sonographer, the number of pulses per second that are transmitted by the transducer must decrease. This occurs because the transducer must wait a longer time for echoes to return from deeper depths. Sonographers therefore adjust the PRF when they adjust the maximum imaging depth during an exam.

114. **B**. The pulse repetition frequency is the number of pulses produced by the ultrasound system in one second. PRF has the same units as frequency: Hertz, Hz, per second, $\sec^{-1}$, or 1/second.

115. **A**. The ultrasound system, the source of the acoustic wave, is the sole determinant of the pulse repetition frequency. The medium through which the sound travels does not directly impact the PRF.

116. **B**. When the maximum imaging depth is increased, the ultrasound machine must wait and listen a longer time for reflections. The longer listening time reduces the system's ability to send out many pulses per second. Hence, as imaging depth increases, the pulse repetition frequency decreases.

117. The pulse repetition frequency is the _____.
   a) product of the wavelength and propagation speed
   b) reciprocal of the period
   c) sum of the pulse duration and the listening time
   d) reciprocal of the pulse repetition period

Two ultrasound systems, one producing sound with frequency of 3 MHz and
the other at 6 MHz, are used to image a patient. The maximum imaging
depth of both exams is 8 cm. Given these facts, are the following four
statements true or false?

118. The pulses produced by both systems travel at the same speed in the
   patient.

119. The PRF of the 6 MHz transducer is greater than the PRF of the 3 MHz
   transducer.

120. The period of the 3 MHz sound is greater than the period of the 6 MHz
   sound.

121. The wavelength of the 3 MHz ultrasound is greater than that of the 6 MHz
   sound.

122. The pulse repetition period is dependent on _____.
   a) the source of the sound wave
   b) the medium that the pulse travels in
   c) both a and b
   d) neither a nor b

123. A sonographer adjusts the output power of the wave emitted by the
   ultrasound transducer. Which of the following also changes? (More than
   one answer may be correct.)
   a) pulse repetition period   b) PRF
   c) propagation speed   d) intensity

124. A sonographer adjusts the maximum imaging depth of an ultrasound
   system. Which of the following also changes? (More than one answer
   may be correct.)
   a) pulse repetition period   b) wavelength
   c) pulse repetition frequency   d) frequency

117. **D.** The pulse repetition frequency and the pulse repetition period are reciprocals. For example, if the PRF is 100 per second, then the pulse repetition period is one-hundredth of a second. If there are fifty pulses per second, then the pulse repetition period is one-fiftieth of a second.

118. **True.** When sound waves travel in the same medium, they travel at exactly the same speed. Their frequencies are irrelevant.

119. **False.** The pulse repetition frequency of an imaging exam is derived from the maximum imaging depth as established by the sonographer. *The PRF changes only when the imaging depth changes.* Because the maximum imaging depth for both systems is the same, the PRFs are the same.

120. **True.** The period of a wave is the length of time required to complete *one single cycle*. A sound signal's period and frequency are reciprocals; therefore, the lower the frequency, the higher the period. In this case, the 3 MHz wave has a longer period than a 6 MHz wave.

121. **True.** Wavelength is the distance that a single cycle of a pulse occupies in space. When traveling through a particular medium, cycles from waves with lower frequencies have longer wavelengths. Thus, cycles from a 3 MHz wave have longer wavelengths than those from a 6 MHz wave.

122. **A.** Similar to the PRF, the pulse repetition period depends only on the source of the wave, namely the ultrasound system. The pulse repetition period is the time from the start of one pulse to the start of the next pulse. The two components of the pulse repetition period are the pulse duration and the duration of which the machine listens for returning echoes. When the sonographer adjusts the imaging depth for an exam, the pulse repetition period is altered. To be more exact, the listening time of the transducer is lengthened when the imaging depth is increased, and it is shortened when the maximum depth is decreased.

123. **D.** As a sonographer adjusts the output power, the intensity will change. On the other hand, the PRF and pulse repetition period will only change when the maximum imaging depth (also called depth of view) changes. The sound's propagation speed will change only if the medium changes.

124. **A and C.** As the imaging depth is altered, the number of pulses that the system produces in 1 second (the pulse repetition frequency) changes. The pulse repetition period is also altered since the PRF and pulse repetition period are reciprocals. The wavelength and frequency remain constant because these terms describe the attributes of a single cycle within the pulse and are not affected by alterations in imaging depth.

125. Which of the following correctly describes the pulse repetition period?
     a) the product of wavelength and propagation speed
     b) the reciprocal of the frequency
     c) the sum of the pulse's "on" time and the listening "off" time
     d) the time during which the transducer is pulsing

126. What happens to the pulse repetition period if the sonographer decreases the maximum imaging depth achieved in an ultrasound scan?
     a) increases   b) decreases
     c) remains the same   d) it cannot be determined

127. What are the units of pulse duration?
     a) units of frequency (Hz, etc.)   b) msec only
     c) units of time (sec, years, etc.) d) units of distance (feet, etc.)

128. In diagnostic imaging, the pulse duration is determined by the _____.
     a) source of the wave   b) medium through which the pulse travels
     c) both a and b   d) neither a nor b

129. What happens to the pulse duration when a sonographer decreases the maximum imaging depth achieved in an ultrasound scan?
     a) increases   b) decreases
     c) remains the same   d) it cannot be determined

130. The characteristics of a pulse are as follows: the pulse repetition period is 1000 µsec, and the listening (or dead) time is 950 µsec. What is the pulse duration?
     a) 1950 µsec   b) 50 msec
     c) 50 µsec   d) 0.95

Are the following two statements regarding the pulse duration true or false?

131. The duration of an acoustic pulse can be adjusted by a sonographer because it is dependent upon the pulse's propagation speed.

132. The duration of an acoustic pulse can be adjusted by the sonographer because it is dependent upon the maximum imaging depth.

125. **C.** The pulse repetition period is the time from the start of one pulse to the start of the next pulse. It is equal to the time that the transducer is pulsing (the pulse duration) plus the time that the ultrasound system is listening for reflected echoes.

126. **B.** As the maximum imaging depth is decreased, the interval of time that the ultrasound machine waits and listens for returning echoes is diminished. As a result of this shorter listening time, the pulse repetition period is shortened.

127. **C.** The pulse duration is the span of time required for the transducer to create a pulse. Hence, it has units of time. The typical range of pulse durations found in diagnostic imaging equipment is 0.4 to 4 μsec; it is, however, valid to report pulse duration in any unit of time.

128. **A.** The pulse duration is the span of time required for the transducer to create an acoustic pulse. It *does not* include the listening time! The pulse duration is a property of the ultrasound system, and it cannot be changed. Each transducer and machine combination produces a characteristic pulse with a fixed duration. The pulse duration remains constant as long as the system functions properly.

129. **C.** The "pulsing" time of the transducer cannot be changed by anything that the sonographer does. The only thing that can change the pulse duration is damage to the transducer or the ultrasound system.

130. **C.** A certain length of time is required for the transducer to transmit and listen. The transmission time is called the pulse duration. The sum of the transmitting and listening times is called the pulse repetition period. If the pulse repetition period is 1000 μsec and the receiving time is 950 μsec, then the transmitting time must be 50 μsec .

131. **False.** The pulse duration cannot be changed by a sonographer. It is a fixed feature, or characteristic, of the ultrasound system. It does not depend upon propagation speed.

132. **False.** The pulse duration cannot be changed by the sonographer. It has a constant value, and it is not dependent on imaging depth.

Are the following two statements regarding the pulse duration true or false?

133. The duration of an acoustic pulse cannot be changed by the sonographer unless he or she switches transducers.

134. The duration of an acoustic pulse cannot be changed under any circumstances or by any action of the sonographer.

135. What is the pulse duration equal to?
    a) frequency multiplied by period
    b) period multiplied by wavelength
    c) the number of cycles in the pulse divided by the wavelength
    d) period multiplied by the number of cycles in the pulse

136. Two pulses are produced by different transducers. The pulses are made up of the same number of cycles. The pulse containing cycles of a lower frequency has a_____ .
    a) lower pulse repetition frequency   b) shorter pulse duration
    c) longer pulse duration   d) longer pulse repetition period

137. The pulse duration is expressed in the same units as the _____ .
    a) period   b) PRF
    c) wavelength   d) density

138. The spatial pulse length describes certain characteristics of an ultrasound pulse. What are its units?
    a) time   b) Hertz
    c) meters   d) none; it is unitless

139. A typical value for the duty factor (also called the duty cycle) of a pulsed ultrasound wave used in diagnostic imaging is _____ .
    a) 0.001 μsec   b) 0.001 kg/cm^3
    c) 0.75   d) 0.001

140. In the case of pulsed ultrasound, what is the maximum value of the duty factor ?
    a) equal to 100   b) equal to 1
    c) less than 100%   d) none of the above

133. **True**. The pulse duration depends upon the interaction of the pulser electronics of the machine and the transducer. If the sonographer does change transducers, then the pulse duration may change.

134. **False**. It is possible for a sonographer to alter the pulse duration by using a different imaging transducer or ultrasound machine. Under these conditions, the pulse duration may be modified.

135. **D**. The pulse duration is the total time span required for the transducer to produce a pulse. The pulse duration is equal to the time to make a single cycle (the period) multiplied by the number of cycles that make up the pulse. For example, if there are 6 cycles in a pulse, each with a period of 2 μsec, then the pulse duration is 6 x 2 or 12 μsec.

136. **C**. The pulse duration is calculated by multiplying the number of cycles in a pulse by the period of a each cycle. When the frequency of the wave is lower, a longer period results. When the period is longer and the number of cycles is the same, the pulse duration is also longer.

137. **A**. The pulse duration is a time span and is measured in units such as seconds, minutes, or hours. The period is also measured in units of time. PRF has units of hertz. Wavelength has units of distance. Density has units of mass per volume.

138. **C**. The spatial pulse length is the length that a pulse occupies in space. Its length is measured from the beginning to the end of the pulse. It can be reported in any units of distance.

139. **D**. Duty factor, or duty cycle, is the *percentage* of time that an ultrasound system is producing an acoustic signal, or transmitting. Typically, ultrasound transducers spend the vast majority of time receiving and only a small percentage of time transmitting an acoustic signal.

140. **C**. This question is a bit tricky. When dealing with a pulsed ultrasound system, at least a tiny bit of time must be spent listening rather than transmitting. Therefore, the percentage of time spent transmitting must be lower than 1 or 100%. If the percentage equals 100%, the system is continuous wave. The correct answer, C, distinguishes between pulsed and continuous wave by stating that the maximum value of the duty cycle must be *less than* 100% for a pulsed system.

141. What is the value of the duty cycle for continuous wave ultrasound?
    a) 100   b) 1%
    c) 1000%   d) none of the above

142. When a particular imaging system is used, what happens to the duty cycle
    when the maximum imaging depth increases?
    a) increases   b) decreases

143. True or False? The instrument operator alters the duty cycle when
    adjusting the maximum imaging depth of a scan.

144. True or False? The duty cycle is a characteristic of an ultrasound and
    transducer system and does not change as long as the system components
    remain unchanged.

145. True or False? The pulse duration of an ultrasound and transducer system
    does not change significantly as long as the system components remain
    unchanged.

146. What is the duty factor of a wave that has a pulse repetition period of 30
    microseconds and a pulse duration of 0.3 microseconds?
    a) 0.03   b) 0.90
    c) 30.3   d) 0.01

The maximum imaging depth obtained during an exam is unchanged. A new
    transducer with a longer pulse duration is used. Given these facts, are the
    following two statements true or false?

147. The pulse repetition period is increased.

148. The pulse repetition frequency is increased.

141. **D**. The duty factor for continuous wave ultrasound is 1.0 or 100%. This means that the transducer is producing an acoustic signal at all times. None of the answers indicate this: 100 is not 100%, 1% means one-hundredth and is incorrect, and 1000% means 10 times.

142. **B**. The duty factor is the *percentage* of the time that an imaging system is producing a pulse. It is equal to the pulse duration divided by the pulse repetition period. Under normal operation, the pulse duration never changes. However, as the maximum imaging depth increases, the pulse repetition period increases. Thus, the duty cycle decreases as imaging depth increases. As the system images to a greater depth, the system spends more time listening for reflections, and the duty factor decreases.

143. **True**. The *percentage* of time required for an ultrasound system to create an acoustic wave is called the duty cycle. The shallower the maximum imaging depth, the higher the percentage of time the machine is emitting sound. The percentage of time for sound transmission is less when the system is imaging to a greater depth.

144. **False**. The duty cycle changes as the maximum imaging depth is altered during an exam by the sonographer.

145. **True**. The pulse duration is the timespan in which a pulse exists. It is determined by the ultrasound system and the transducer. It never changes unless something is drastically wrong, such as a transducer failure or electronic problems in the machine.

146. **D**. The duty cycle is calculated by dividing the pulse duration by the pulse repetition period. The pulse duration is 0.3 msec and the pulse repetition period is 30 msec. 0.3 divided by 30 is 0.01. This is the duty cycle.

147. **False**. The pulse repetition period is determined by the maximum imaging depth achieved in an exam. In this case, the maximum depth is unchanged and; therefore, the pulse repetition period is also unchanged.

148. **False**. Pulse repetition frequency and pulse repetition period are reciprocals. If the pulse repetition period is unchanged, then the PRF must also remain unchanged.

The maximum imaging depth obtained during an exam is unchanged. A new transducer with a longer pulse duration is used. Given these facts, are the following two statements true or false?

149. The duty factor is increased.

150. The frequency is increased.

151. While using the same ultrasound machine and transducer, which of the following can a sonographer alter? (More than one answer may be correct.)
    a) pulse repetition period   b) PRF   c) frequency
    d) duty cycle   e) pulse duration

152. In diagnostic imaging, what determines the spatial pulse length?
    a) the ultrasound system   b) the medium through which the pulse travels
    c) both a and b   d) neither a nor b

153. Which of the following best describes the spatial pulse length?
    a) frequency multiplied by wavelength
    b) PRF multiplied by wavelength
    c) wavelength multiplied by the number of cycles in the pulse
    d) duty factor multiplied by the wavelength

154. Two transducers send ultrasound pulses into soft tissue. One transducer emits sound with a 4 MHz frequency, and the other emits sound at 6 MHz frequency. Each pulse comprises 4 cycles. Which has a greater spatial pulse length?
    a) the 6 MHz pulse   b) the 4 MHz pulse
    c) they are the same   d) it cannot be determined

155. Using a given transducer and machine, what happens to the spatial pulse length as the sonographer increases the maximum imaging depth?
    a) increases   b) decreases
    c) remains the same   d) it cannot be determined

156. True or False? While imaging soft tissue, the spatial pulse length does not change as long as the components of the ultrasound system are the same.

149. **True**. The duty cycle is calculated by dividing the pulse duration by the pulse repetition period. In this example, the pulse repetition period is unchanged. By changing to a transducer with a longer pulse duration while the pulse repetition period stays constant, the duty factor increases.

150. **False**. From the information given in the question, nothing is stated regarding the frequency of the signal emitted by the transducer. Therefore, nothing definite can be concluded.

151. **A, B, and D**. While adjusting the desired maximum imaging depth in an exam, the sonographer adjusts the pulse repetition period and the PRF. In addition, the duty factor (the *percentage* of the transmission time of the transducer) is altered. Unlike choices A, B, and D, the frequency of the ultrasound and the pulse duration are fixed once a transducer is selected. The sonographer cannot alter these parameters.

152. **C**. The spatial pulse length is the distance, or length, of a pulse. It depends, in part, upon the wavelength of each individual cycle in the pulse. Wavelength depends upon both the source of the sound and the medium through which the sound travels. Similarly, the length of the entire pulse also depends upon both the source and the medium.

153. **C**. The total length of a pulse equals the length of each cycle in the pulse multiplied by the number of cycles in the pulse. Imagine the pulse as a train comprising several boxcars. The train's length (the spatial pulse length) equals the length of each car (the wavelength) multiplied by the number of cars in the train (the number of cycles in the pulse).

154. **B**. Since both pulses have the same number of cycles, the pulse comprising individual cycles of longer wavelengths will have the greater overall length. In a given medium, waves with lower frequencies have longer wavelengths. The 4 MHz wave has a longer wavelength than the 6 MHz wave; therefore it will have a longer spatial pulse length.

155. **C**. The spatial pulse length is determined by the number of cycles in the pulse and the wavelength of each cycle. Within the same system, these factors are unchanged, and the spatial pulse length remains the same.

156. **True**. The overall length of a pulse is equal to the wavelength multiplied by the number of cycles contained in the pulse. Within a particular ultrasound system and transducer, the pulse length cannot change.

157. The propagation speed of continuous wave ultrasound is 1.8 kilometers per second. The wave is then pulsed with a duty factor of 0.5. What is the new propagation speed?
    a) 0.5 km/sec   b) 0.9 km/sec
    c) 1.8 km/sec   d) 3.6 km/sec e) cannot be determined

158. The frequency of a continuous acoustic wave is 5 MHz. The wave is then pulsed with a duty factor of 0.1. What is the new frequency?
    a) 0.5   b) 0.5 MHz
    c) 5 MHz   d) 10 MHz

159. True or False? The period of an ultrasound wave is related to the frequency of the acoustic signal, and it is the same regardless of whether the wave is pulsed or continuous.

160. True or False? The wavelength of an acoustic wave is shorter when it is pulsed rather than continuous.

161. The term I_m defines the _____ and has units of

    _____ .
    a) medium's maximum impedance, Rayls
    b) transducer's minimum input, watts
    c) maximum intensity, watts/cm^2
    d) minimum inductance, Rayls/sec

162. With a continuous wave sound beam, which of the following four intensities are the same?
            1) SPTA   2) SATA   3) SATP   4) I_m

    a) 1 and 2   b) 1 and 3
    c) 1 and 4   d) 2 and 3

163. The SPTP intensity of a typical pulsed acoustic wave _____ .
    a) exceeds the SATA intensity   b) exceeds the SPTA intensity
    c) exceeds the SATP intensity   d) all of the above

157. **C.** The speed of sound in a medium is determined only by the medium. There is no difference in the speed of sound between continuous and pulsed waves. Thus, the new and old speeds will be identical: 1.8 km/sec.

158. **C.** Frequency is the reciprocal of the period. Another way to consider frequency is: if the wave were continuous, how many cycles of the acoustic variable would occur in one second? Therefore, the fact that a wave is pulsed, rather than continuous, does not alter the frequency of the signal. The new and old frequencies are the same: 5 million/second.

159. **True.** The period of a wave is a characteristic of each individual cycle in the acoustic signal. Therefore the period of a continuous or pulsed wave (as well as frequency and wavelength) does not change.

160. **False.** The wavelength is a parameter that reports the length of a single cycle in a wave. Therefore, the wavelength does not change, whether it is part of a pulsed or continuous wave.

161. **C.** The term I_m means the maximum intensity that is observed in an ultrasonic wave when averaged over the largest one-half cycle. As is true of all intensities, I_m is power divided by beam area. The correct units are watts per centimeter squared: W/cm^2.

162. **D.** With a continuous wave, temporal peak and temporal average intensities are the same. The term temporal means dealing with "time." The temporal terms are identical for continuous wave because the beam is always "on." Thus, the SATA (spatial average, temporal average) and SATP (spatial average, temporal peak) intensities have the same value.

163. **D.** With a pulsed wave, the SPTP is the largest intensity. When an acoustic wave is pulsed, the temporal peak always exceeds the temporal average. In addition, in a typical acoustic wave, the intensity near its center is greater than that of its edges. Therefore, the spatial peak intensity is higher than the spatial average intensity. This leads to the conclusion that for pulsed waves, the SPTP intensity has the highest value.

With regard to the SPPA intensity, are the following 4 statements true or false?

164. It has a value that is greater than the SPTP intensity.

165. It is only relevant for continuous wave ultrasound.

166. It has a value that is between the SPTP and the SPTA intensities.

167. It can be reported in units of watts per square centimeter.

168.   What does the beam uniformity coefficient measure?
    a) special distribution of sound energy
    b) spatial distribution of acoustic energy
    c) temporal distribution of sound energy
    d) none of the above

169. Which of the following values can correctly designate both the duty cycle
    and the beam uniformity coefficient?
    a) 1%   b) 0
    c) 100%   d) none of the above

170. Which of the following intensities changes when a sonographer adjusts the
    maximum imaging depth obtained during a sonographic examination?
    a) SPTP   b) SATA
    c) SATP   d) SAPA

171. What are acceptable units for the beam uniformity coefficient?
    a) mW   b) mW/cm^2
    c) cm   d) none of the above

164. **False**. The spatial peak pulse average (SPPA) intensity is less than the spatial peak temporal peak (SPTP) intensity. The intensity of a pulse produced by an ultrasound transducer is not constant and varies throughout the duration of the pulse. The SPTP is the greatest intensity at any time, while the SPPA is the average intensity during the pulse duration.

165. **False**. The SPPA (spatial peak pulse average) intensity is relevant for pulsed ultrasound and is meaningless for continuous wave US. Since no "pulse" exists in continuous wave US, the term SPPA does not apply.

166. **True**. The largest intensity is the SPTP. Of the choices provided, the SPTA is the smallest intensity . The SPPA is the average intensity only during the pulse duration; it has a value between the SPTP and SPTA.

167. **True**. The SPPA is an intensity of an acoustic wave. Intensity, regardless of what type, always has units of power per unit area, or in this case, watts per square centimeter.

168. **B**. The term spatial, meaning "throughout space," is related to the beam uniformity coefficient. The beam uniformity coefficient is an index of how evenly an ultrasound beam is distributed throughout space.

169. **C**. The beam uniformity coefficient is equal to or greater than one. The duty cycle ranges from zero to one. The only number common to both the beam uniformity coefficient and the duty factor is one. 100% is another way of expressing the value one.

170. **B**. When a sonographer alters the maximum imaging depth, the receiving time of the system is adjusted. The SPTP and SATP are measures of intensities at specific *times* when the intensities are maximal. They are not affected by the receiving time (when the intensity is zero). The SAPA is the average intensity when the transducer is "pulsing" and it also ignores the receiving time. The only intensity that is affected by both the listening and pulsing times is the SATA.

171. **D**. The beam uniformity coefficient is a *unitless* number that indicates the degree to which an ultrasound beam's intensity is distributed in space. It is calculated by dividing the spatial peak by the spatial average intensity.

172. Two acoustic beams have identical SPTP intensities of 400 mW/cm^2. One beam is pulsed, and the other is continuous wave. Which beam has a higher SPTA intensity?
     a) the pulsed beam   b) the continuous wave
     c) neither   d) it cannot be determined

173. Two acoustic beams have the same SPTP intensities. One beam is produced by a continuous wave transducer, whereas the other comes from a pulsed wave system. Which wave has the higher SATP intensity?
     a) the pulsed wave   b) the continuous wave
     c) neither, both are the same d) it cannot be determined

174. True or False? The SPTA intensity exceeds the SPPA intensity for pulsed acoustic waves.

175. True or False? The SPTP intensity is never less than the SPTA intensity for all types of acoustic waves.

176. What is another name for the beam uniformity coefficient?
     a) duty cycle   b) attenuation coefficient
     c) SP/SA factor   d) beam impedance ratio

177. The logarithm of a numeral is defined the number of times _____ must be multiplied together to get that numeral.
     a) 1   b) 2
     c) 5   d) 10

Match the following five numbers with their logarithms. (Each number in the left hand column has a match in the right hand column).

178. 5   a) 1.0
179. 10   b) 2.0
180. 40   c) 0.7
181. 100   d) 1.6
182. 317   e) 2.5

172. **B.** With continuous wave ultrasound, the SPTP and SPTA intensities are the same: 400 mW/cm^2. On the other hand, with pulsed wave US, the SPTP always exceeds the SPTA intensity because there is a time when the transducer is not producing a wave and is listening for reflected echoes. With pulsed US, the SPTA intensity is less than 400 mW/cm^2, and the continuous wave will have the higher SPTA intensity.

173. **D.** There is insufficient information to determine if the pulsed or continuous wave has the higher SATP intensity. This question can't be answered unless the beam uniformity coefficients for the waves are known.

174. **False.** Without exception, the SPTA (spatial peak temporal average) intensity is always less than the SPPA (spatial peak pulse average) intensity. The SPTA intensity includes the time when the transducer is not producing a pulse and is "listening" for reflected echoes. The SPPA intensity includes only the transmitting time of the transducer. Therefore, the SPPA intensity will always exceed the SPTA intensity.

175. **True.** The SPTP (spatial peak temporal peak) intensity is never smaller than the SPTA (spatial peak temporal average) intensity. The SPTP intensity is the largest intensity recorded at any position in space and at any point in time. Therefore, SPTP is the largest of all intensities.

176. **C.** The beam uniformity coefficient is also called the SP/SA factor.

177. **D.** The logarithm of a numeral is equal to the number of TENS that are multiplied together to arrive at that number.

NOTE: With this group of five questions, the logarithms of specific numbers are unimportant. We should know, however, that as a number gets larger, so does its logarithm. In this sequence, the five numbers in ascending order are to be matched with their logarithms. This is most easily achieved by also ranking the logarithms in increasing order.

178. **C.** The logarithm of 5 is equal to 0.7.

179. **A.** The logarithm of 10 is equal to 1.0.

180. **D.** The logarithm of 40 is equal to 1.6.

181. **B.** The logarithm of 100 is equal to 2.0.

182. **E.** The logarithm of 317 is 2.5.

183. What is a decibel?
    a) the absolute value of a number   b) a range of values
    c) a relationship between two numbers   d) none of the above

184. The intensity of a signal decreases from 1.5 mW/cm^2 to 0.75 mW/cm^2.
    How many decibels is this change in intensity?
    a) 3 dB   b) 0.75 dB
    c) -0.75 dB   d) -3 dB

185. What is the decibel representation when an acoustic signal is amplified?
    a) positive   b) negative
    c) equal to zero

186. What is the decibel notation for an acoustic signal that is attenuated?
    a) greater than zero   b) less than zero
    c) equal to zero

187. The power in a wave is increased to ten times its original value. How
    many decibels are needed to illustrate this change?
    a) 3   b) 6
    c) 10   d) 20

188. How many decibels represent a 100-fold increase in the intensity of an
    acoustic pulse?
    a) 10   b) 20
    c) 100   d) 6

189. The intensity of an ultrasound wave is changed by -6 dB. This means that
    the current intensity is _____ as much as its original level.
    a) one-tenth   b) four times
    c) one-fourth   d) ten times

190. The scale associated with the decibel notation is _____.
    a) linear   b) discrete
    c) logarithmic   d) additive

191. What does a 3 dB change in a value of a parameter mean?
    a) the value has doubled   b) the value has tripled
    c) the value has increased 30% d) the value has increased ten times

183. **C.** A decibel represents a relationship between two numbers. A decibel is a relative measure of intensity or power. The term "relative" indicates that we are not dealing with an absolute power (such as 600 watts), but rather how the power is related to a reference level (e.g., the current signal is 50% as strong as the reference level).

184. **D.** A decline in intensity to one-half of the original level is a change of -3 dB. When the new level is less than the reference level, the reported decibels must be negative. When the new intensity is greater than the reference level, the change in decibels is positive.

185. **A.** When a signal is amplified, its magnitude is increasing and it becomes stronger. In decibel notation, an increase in signal strength is described with positive decibels.

186. **B.** When a signal is attenuated, its magnitude is decreasing and it becomes weaker. In decibel notation, a decrease in signal strength is described with negative decibels.

187. **C.** An increase of ten times the original power of a wave is reported as +10 dB.

188. **B.** An increase of 100 times the intensity of a wave is equal to 20 dB. Each 10 dB indicates a tenfold increase. Therefore, 20 dB (two 10-dB increases) means a pair of tenfold increases: 10 x 10 = 100.

189. **C.** A change in intensity of -6 dB indicates that only one-fourth of the initial intensity remains. The minus sign indicates a decrease in signal magnitude. Each -3 dB change means that one-half of the original intensity remains. Since there are two sets of -3 dB; one-half multiplied by one-half indicates that there is only one-fourth remaining.

190. **C.** Decibel notation is based on the mathematical relationship of logarithms.

191. **A.** A change in the value of a parameter of 3 dB indicates that the value has doubled.

192. As sound propagates through a medium, the total power in the wave decreases. What is this entire process called?
    a) absorption
    b) scattering
    c) attenuation
    d) reflection

193. An acoustic wave is traveling through soft tissue. Its intensity undergoes six decibels of attenuation. How does the final intensity of the wave relate to the intensity of the wave when it started its journey?
    a) it is now four times larger
    b) it is now six times larger
    c) it is now one-fourth as large
    d) it is now one-tenth as large

194. Which of the following is not one of the physical processes that contribute to attenuation of ultrasound waves passing through soft tissue?
    a) reflection
    b) redirection of sound in many directions
    c) focusing
    d) conversion of acoustic energy to heat

195. The attenuation of an acoustic wave traveling through bone is _____ its attenuation through soft tissue.
    a) greater than
    b) less than
    c) equal to

196. Ultrasound waves traveling through lung tissue attenuate to a _____ extent than when traveling through soft tissue.
    a) greater
    b) lesser
    c) nearly equal

197. Sound traveling through blood attenuates to a _____ extent than when traveling through soft tissue.
    a) greater
    b) lesser
    c) relatively equal

198. As a pulse passes through soft tissue, a certain amount of acoustic energy remains in the tissue as heat. What is this constituent of attenuation called?
    a) scattering
    b) absorption
    c) refraction
    d) rarefaction

192. **C.** As a wave propagates through a medium, its power diminishes. This is attenuation. Choices A, B and D are all components of or contributors to attenuation.

193. **C.** When a wave undergoes 6 dB of attenuation, the intensity of the wave is decreased to one-fourth of its initial value. The term attenuation means "to weaken or reduce in force or intensity." Therefore, any time the term attenuation is used, the intensity of the wave must be diminishing. Six decibels of attenuation are made up of two groups of -3 dB. Each -3 dB indicates a halving of intensity. One-half multiplied by one-half means that only one-fourth of the original intensity remains.

194. **C.** Focusing does not contribute to the attenuation process. In contrast, choices A, B, and D all cause a reduction in the intensity of a wave as it propagates.

195. **A.** The attenuation of ultrasound in bone is greater than its attenuation in soft tissue.

196. **A.** Ultrasound attenuates more when traveling in lung than in soft tissue.

197. **C.** Ultrasound waves attenuate at similar rates in both media.

198. **B.** The conversion of acoustic energy into heat is called absorption. Through this process, a portion of the energy from the beam is deposited in the soft tissues.

199. What is the formal name for dispersion of a wave in many different
     directions after striking a small particle?
     a) microscattering   b) backscattering
     c) Rayleigh scattering   d) total absorption

200. Which of the following is considered a Rayleigh scatterer?
     a) bone   b) liver
     c) muscle   d) blood

201. A sound wave reaches a border between two media. The wavelength of
     the acoustic wave is smaller than the irregularities in the boundary. Under
     these explicit circumstances, which process is most likely to occur?
     a) backscatter reflection   b) specular reflection
     c) Rayleigh scattering   d) refraction

202. An acoustic pulse reflects from a boundary where the irregularities on the
     surface of the boundary are much larger than the pulse's wavelength.
     What type of reflection is most likely to occur under these circumstances?
     a) partial   b) Rayleigh
     c) specular   d) total

203. The direction of motion of a particle in a wave is perpendicular to the
     direction of propagation of the wave. What type of wave is this?
     a) longitudinal   b) acoustic
     c) mechanical   d) transverse

204. A sound wave strikes a boundary between two media at a 60° angle. This
     is called _____ incidence.
     a) orthogonal   b) angular
     c) obtuse   d) oblique

205. Which of the following describes an angle with a measurement of 45°?
     a) orthogonal   b) acute
     c) obtuse   d) normal

206. Which of the following describes an angle with a measurement of 123°?
     a) orthogonal   b) acute
     c) obtuse   d) normal

199. **C.** Rayleigh scattering occurs when an acoustic wave scatters in many different directions after striking a small particle. When Rayleigh scattering occurs, the dimension of the reflecting particle is usually less than the wavelength of the ultrasound wave. The amount of Rayleigh scattering also depends on the frequency of the ultrasound.

200. **D.** Blood is a Rayleigh scatterer. A red blood cell is smaller than the wavelength of the typical acoustic wave used in diagnostic imaging. When an acoustic wave strikes a blood cell, the energy within the pulse is scattered in many directions.

201. **A.** The smoothness or roughness of a boundary will help to determine what form of reflection will take place. Backscatter is likely to occur when the boundary has irregularities that are larger than the wavelength of the incident acoustic pulse.

202. **C.** A boundary is considered smooth when irregularities in its surface are larger than the wavelength of the incident ultrasonic beam. Reflections from a smooth boundary are specular. A mirror is a specular reflector. The waves strike this smooth boundary and reflect in an organized, systematic manner.

203. **D.** A transverse wave has the characteristic that the direction of propagation of the wave is perpendicular to the direction of particle motion in the wave. Sound waves are not transverse, but rather are longitudinal.

204. **D.** Oblique incidence is always present when the angle between the direction of a wave's propagation and the boundary between two media is different from 90°. Oblique incidence is a definition of exclusion. That is, if the incidence is not perpendicular, it is oblique.

205. **B.** Any angle with a measurement less than 90° is an acute angle.

206. **C.** Any angle with a measurement greater than 90° is an obtuse angle.

207. The angle between the direction of propagation and the boundary between two media is exactly 90°. What term appropriately describes the form of incidence of the wave?
    a) not normal   b) direct
    c) oblique   d) orthogonal

208. Which term does not belong in the group below?
    a) orthogonal   b) at right angles   c) oblique
    d) 90° e) normal   f) perpendicular

209. Which term has a meaning other than normal incidence?
    a) orthogonal incidence   b) perpendicular incidence
    c) oblique incidence   d) acute incidence

210. Acoustic impedance is a property of the _____ and has units of _____.
    a) source & medium, Imps   b) medium, dB
    c) medium, Rayls   d) medium, Ohms

211. In order to calculate the acoustic impedance of a medium, one should _____ the _____ by the _____.
    a) divide, propagation speed, density
    b) multiply, density, propagation speed
    c) divide, density, propagation speed
    d) multiply, stiffness, density

212. Which two attributes help to establish the acoustic impedance of a medium?
    a) density and temperature   b) density and stiffness
    c) stiffness and elastance   d) elasticity and compressibility

213. Which value is closest to the impedance of soft tissue?
    a) 1.5 kiloRayls   b) 2.5 deciRayls
    c) 19 megaRayls   d) 2,000,000 Rayls

207. **D.** Orthogonal incidence is attained when a sound wave strikes a boundary between media at exactly 90°. When the incidence is not orthogonal, it is called oblique.

208. **C.** The five terms: orthogonal, at right angles, 90°, normal, and perpendicular are synonymous and their meanings are identical. The term oblique means "not normal."

209. **C.** Oblique incidence is different from normal incidence. Oblique incidence occurs when a wave strikes a boundary at any angle other than 90°, whereas normal incidence occurs when the wave strikes a boundary at *exactly* 90°.

210. **C.** The acoustic impedance is a property of the medium through which sound travels. It is unaffected by the sound source or the characteristics of the wave itself. The units of impedance are Rayls.

211. **B.** The value of acoustic impedance is obtained by multiplying the density of a medium by its propagation speed.

212. **B.** The acoustic impedance may be calculated by multiplying the propagation speed of the medium by the density of the medium. The propagation speed of a medium is determined by the density and the stiffness of the medium. Therefore, acoustic impedance is a property of both the density and the stiffness.

213. **D.** The impedance of soft tissue is in the range of 1.5 to 2.5 million Rayls. Choices A and B are far too low, and choice C is too high.

214. If the _____ of two media are different and sound strikes a border between the media at an orthogonal incidence, then reflection will take place.
   a) conductances
   b) densities
   c) impedances
   d) propagation speeds

215. True or False? The proportion of the incident intensity that is reflected at a border between two media will increase as the impedances of the media become increasingly dissimilar.

216. An ultrasound pulse strikes a boundary between two media with normal incidence. The propagation speeds of sound in the two media are very different; however, the impedances of the media are identical. What will happen?
   a) a big echo will be produced
   b) a medium echo will result
   c) a small echo will result
   d) there will be no echo at all

217. Two acoustic waves strike a boundary between two media. The waves are traveling in a direction 90^0 to the boundary. Reflection of these waves depends on differences in the _____ .
   a) frequencies of the two waves
   b) propagation speeds of the two media
   c) amplitudes of the two waves
   d) impedances of the two media

Two waves, a 5 MHz ultrasonic wave and a 5 kHz audible wave, travel through soft tissue. Are the following four statements true or false?

218. The ultrasound wave travels much faster than the audible wave.

219. Both the 5 MHz and the 5 kHz waves travel at similar speeds through the medium.

220. The wavelength of the 5 MHz wave is greater than the wavelength of the 5 kHz wave.

221. The period of the 5 MHz wave is less than the period of the audible wave.

214. **C.** Under the conditions of orthogonal or normal incidence, reflection depends on differences in the acoustic impedances of the media on either side of the boundary. With normal incidence, as long as the impedances are dissimilar, reflection will always occur.

215. **True.** As a sound waves strikes the border that separates two media, reflection will occur if the impedances of the media are different. The greater the difference between the two impedances, the stronger the resultant reflection. If the impedances are only slightly different, then a weak reflection will be produced.

216. **D.** When a sound wave is normally incident at a boundary, reflection will occur only when the media have dissimilar impedances. In this example, the wave is normally incident but the media have identical impedances. Under these circumstances, it is impossible for any reflection to occur.

217. **D.** With normal incidence, reflection will occur at a boundary when the media on either side of the boundary have different acoustic impedances. Reflection is not dependent on the waves' characteristics such as amplitude and frequency. With normal impedance, there is only one condition that must be met for reflection to take place: the impedances of the media on either side of the boundary must be different.

218. **False.** All sound waves, regardless of their frequencies, travel at exactly the same speed in a particular medium. Both waves propagate at the same speed. In soft tissue, sound propagates at a speed of 1,540 m/sec.

219. **True.** All sound waves travel at exactly the same speed while moving though a particular medium.

220. **False.** When two sound waves travel through the same medium, the higher frequency wave has a shorter wavelength. The wave with the lower frequency has a longer wavelength. In this question, the 5 MHz wave has a higher frequency and therefore has a shorter wavelength.

221. **True.** The period and frequency of any wave are inversely proportional, that is, the higher the frequency of wave, the shorter the period. In this case, the 5 MHz wave has a higher frequency than that of the 5 kHz wave and thus will have a shorter period.

222. An ultrasound wave approaches an interface between two media at a 90° angle. The propagation speeds and the densities of the two media are different. What is correct?
    a) reflection will definitely occur
    b) reflection will definitely not occur
    c) refraction may occur
    d) none of the above

223. An ultrasound wave travels strikes an interface between two media at a 90° angle. The propagation speeds of the media are identical. However, the densities of the media are different. Which is true?
    a) reflection will definitely occur
    b) reflection will definitely not occur
    c) refraction may occur
    d) none of the above

224. What are the units of the intensity transmission coefficient?
    a) watts/square cm   b) watts
    c) dB   d) none of the above

225. What are the units of the intensity reflection coefficient?
    a) none   b) W/cm^2
    c) watts   d) dB

226. What are the units of the transmitted intensity of a sound wave?
    a) W/cm^2 b) watts
    c) dB   d) none; it is unitless

222. **D.** The facts presented indicate that the propagation speeds and the densities of the two media are different, but this information is not sufficient to determine whether reflection will take place. The important fact is whether the impedances of the media are different. If so, reflection will occur. If the impedances are the same, then reflection will not occur. In this case, since we are unsure whether the impedances of the media are the same, we are unsure whether choice A or B is correct. Choice C is incorrect because refraction will never occur when a sound wave strikes a boundary at a 90° angle. The only appropriate answer is D.

223. **A.** Reflection will definitely occur. With normal incidence, reflection will occur when the impedances of the media on either side of the boundary are different from each other. In this example, the densities of the media are different, whereas their propagation speeds are the same. Since impedance is propagation speed multiplied by density, it follows that the acoustic impedances of the media bordering the interface are indeed different. Reflection will occur.

224. **D.** The intensity transmission coefficient is defined as the *percentage* of an ultrasound beam's intensity that is transmitted as the sound wave passes through a boundary between two media. The intensity transmission coefficient is a percentage and is reported without units.

225. **A.** The intensity reflection coefficient is defined as the *percentage* or proportion of an ultrasound beam's intensity that is reflected as the sound wave strikes a boundary between two media. The intensity reflection coefficient is reported as a percentage; and therefore it is unitless.

226. **A.** As sound propagates, one of the ways to measure its strength is by recording its intensity. The intensity of the part of a beam that continues to propagate in the forward direction as it reaches a boundary is called the transmitted intensity. It is reported in units of W/cm^2.

227. What results when the intensity transmission coefficient and the intensity reflection coefficient are added together?
    a) incident intensity coefficient   b) acoustic impedance
    c) total intensity   d) 1.0

228. What remains when the reflected intensity is subtracted from the incident intensity?
    a) 1.0   b) incident intensity
    c) transmitted intensity coefficient   d) transmitted intensity

229. A wave strikes an interface between two media, and intensities are measured at the interface. What results when the reflected intensity is divided by the incident intensity?
    a) intensity reflection coefficient   b) intensity transmission coefficient
    c) beam uniformity coefficient   d) none of the above

230. An ultrasound wave strikes a boundary between two media. All intensities are measured directly at the boundary. What results when the transmitted intensity is divided by the reflected intensity?
    a) intensity reflection coefficient   b) intensity transmission coefficient
    c) beam uniformity coefficient   d) none of the above

231. An ultrasound wave strikes an interface between two media. All intensities are measured directly at the borderline. What results when the transmitted intensity is divided by the incident intensity?
    a) intensity reflection coefficient   b) intensity transmission coefficient
    c) beam uniformity coefficient   d) none of the above

232. What is the maximum permissible value for both the intensity reflection coefficient and the intensity transmission coefficient?
    a) 100   b) 1%
    c) 1   d) infinity

233. What is the minimum permissible value for both the intensity reflection coefficient and the intensity transmission coefficient?
    a) different from each other   b) -1
    c) 100%   d) 0

227. **D.** The intensity reflection coefficient and intensity transmission
coefficient, when added together, are equal to one. The intensity
transmission coefficient is the percentage of a sound wave's intensity that
is *transmitted* at a boundary. The intensity reflection coefficient is the
percentage of a sound wave's intensity that is *reflected* at a boundary.
Both of these percentages added together equal 100% (or 1.0).

228. **D.** At any point in time and space, there must be conservation of energy.
In other words, all energy must be accounted for. The energy or
intensity that is present in a wave as it reaches a border between two
media is called the incident intensity. As the wave strikes the boundary,
a portion of the wave's intensity is redirected toward the sound source.
This is called the reflected intensity. The remainder of the energy
continues to propagate and is called the *transmitted intensity*.

229. **A.** Under these circumstances, when we divide the reflected intensity by
the incident intensity, the result is the intensity reflection coefficient.
The intensity reflection coefficient is the percentage of the incident
intensity that is reflected as a wave strikes a boundary between 2 media.

230. **D.** The choices of A, B, and C are all incorrect. When the transmitted
intensity is divided by the reflected intensity, a number is obtained. This
number has no special meaning in ultrasound physics and is not
identified by a special name. It is not the intensity reflection coefficient,
the intensity transmission coefficient, or the beam uniformity coefficient.

231. **B.** Under these circumstances, when the transmitted intensity is divided
by the incident intensity, the result is the intensity transmission
coefficient. The intensity transmission coefficient is the percentage of
the incident intensity that continues to propagate in the forward direction
when the incident wave strikes a boundary between two media.

232. **C.** The maximum percentage of the incident intensity that can either be
reflected or transmitted is 1.0 or 100%. At the extremes, total reflection
can occur (intensity reflection coefficient = 1.0), or complete
transmission can occur (intensity transmission coefficient = 1.0).
Therefore, 1.0 is the upper limit of both of these coefficients.

233. **D.** The minimum value for the intensity transmission and the intensity
reflection coefficients is zero. At one extreme, it is possible to have total
transmission and no reflection (intensity reflection coefficient = 0). At
the other extreme, it is possible for total reflection to occur. No
transmission would exist and the intensity transmission coefficient = 0.

234. True or False? When reflection occurs with oblique incidence, the angle
     of reflection equals the angle of incidence. This is known as Snell's Law.

235. Snell's Law is a declaration concerning the behavior of waves during a
     particular physical process or event. What event is this?
     a) reflection   b) transmission
     c) refraction   d) impedance

236. What conditions are necessary for refraction to occur at a boundary
     between two media?
     a) unequal acoustic impedances and normal incidence at the boundary
     b) unequal densities of the media and normal incidence at the boundary
     c) dissimilar propagation speeds and oblique incidence at the boundary
     d) different elasticities of the media and oblique incidence

237. True or False? Refraction occurs at the border between two media only if
     there is oblique incidence of the wave upon the boundary.

238. True or False? Refraction always occurs at the interface between two
     media when the propagation speeds of the media are unequal.

239. An acoustic wave is traveling from medium X into medium Z. Medium X
     has a propagation speed of 1,457 m/sec and an impedance of 1.44
     MRayls. Medium Z has a propagation speed of 1,644 m/sec and an
     impedance of 1.26 MRayls. The angle of incidence is 32°. What is true
     of the angle of the transmitted wave?
     a) it is greater than 32° b) it is equal to 32°
     c) it is less than 32° d) it cannot be determined

240. An acoustic wave in medium B is traveling toward medium A. The sound
     beam strikes the boundary at a 45° angle. The propagation speed is 1,547
     m/sec for medium A and 1,745 m/sec for medium B. What is true of the
     angle of transmission?
     a) $> 45^O$ b) $< 45^O$
     c) $= 45^O$ d) it cannot be determined

234. **False.** It is indeed true that the angle of reflection is equal to the angle of incidence when reflection occurs at an oblique incident angle. The falsehood in this statement is that this is *not* Snell's Law.

235. **C.** Snell's Law defines the physics of refraction.

236. **C.** The conditions required for refraction are twofold: 1) a sound wave must be obliquely incident to the border between two media, and 2) the media on either side of the border must have dissimilar propagation speeds. Refraction occurs during transmission of a wave from one medium to another.

237. **True.** Refraction cannot occur if a wave is normally incident to the boundary between two media. It can occur only if the incident wave is oblique to the boundary.

238. **False.** A sound wave refracts at the boundary between two media when two conditions are met. One of the requirements involves the incidence of the sound beam: it must be obliquely incident to the boundary between the two media. The second requirement is that the propagation speeds of the media on either side of the borderline must be different from each other. The statement is false because it does not account for the possibility that refraction cannot occur when the sound wave is normally incident to the boundary.

239. **A.** With refraction, the relationship between the angle of incidence and the angle of transmission is defined by Snell's Law. When the propagation speed of the medium that sound is entering exceeds that of the medium sound is in, the angle of transmission is greater than the incident angle.

240. **B.** With refraction, when the medium that the sound is in has a greater propagation speed than that of the medium that the sound is entering, the angle of transmission is less than the angle of incidence. In this case, the angle of transmission will be less than 45° because the sound wave is entering a slower medium than that through which it is propagating.

241. True or False. The characteristics of a sound wave determine whether refraction will occur.

242. An acoustic wave is traveling from medium X to medium Z. Medium X has a propagation speed of 1,457 m/sec and an impedance of 1.44 MRayls. Medium Z's propagation speed is 1,644 m/sec, and its impedance is 1.26 MRayls. The angle of incidence is 32^0. What is true of the angle of reflection?
   a) it is greater than 32^0 b) it is equal to 32^0
   c) it is less than 32^0 d) it cannot be determined

243. An acoustic wave in medium A and is traveling toward medium B. The sound beam's angle of transmission into medium B relative to the boundary is 79^0. Sound's propagation speed is 1,547 m/sec in medium A and 1,745 m/sec in medium B. If reflection and transmission both occur at the boundary, what can be said of the reflection angle?
   a) $> 79^0$ b) $< 79^0$
   c) $= 79^0$ d) it cannot be determined

244. A wave made of acoustic energy is leaving soft tissue and proceeding toward fat tissue with an 86^0 incident angle. What is true of the angle of transmission?
   a) 86^0 b) less than 86^0
   c) greater than 86^0 d) it cannot be determined

245. Sound propagates from one medium with a density of 1.16 kg/m^3 to a second medium with a density of 1.02 kg/m^3. If the angle of transmission is 49^0, what is correct regarding the angle of incidence?
   a) it is less than 49^0 b) it is equal to 49^0
   c) it is greater than 49^0 d) it cannot be determined

246. True or False? The characteristic impedance of acoustic gel is greater than the matching layer's impedance, but less than the piezoelectric element's impedance.

241. **False.** Refraction depends only on the medium and the angle of incidence. It is independent of the nature of the sound wave.

242. **B.** This is a tricky question. In cases of oblique incidence, the angle of reflection is always equal to the angle of incidence. Simply put, if the incident angle is 32°, then the reflection angle is 32°. All of the information regarding the propagation speeds and acoustic impedances of media X and media Z are irrelevant in this question.

243. **B.** Two steps are required to determine that the angle of reflection is less than 79°. It is given that the angle of transmission is 79° and that the propagation speed of medium A is less than the propagation speed of medium B. From this, information about the angle of incidence can be derived. Since the speed of medium B is greater than the speed of medium A, the angle of transmission is greater than the angle of incidence. Thus, the angle of incidence must be less than 79°. Second, the angle of reflection is equal to the angle of incidence. Since the angle of incidence is less than 79°, the angle of reflection is also less than 79°.

244. **B.** The propagation speed through fat is less than that of soft tissue. The transmission angle is less than the incidence angle when a wave travels into a second medium that reduces the wave's propagation speed. In this example, the incidence is oblique, and fat has a lower speed than soft tissue. The angle of transmission will be less than the angle of incidence.

245. **D.** No definite statement can be made about the angle of incidence. The angle of incidence depends upon the angle of transmission and the propagation speeds of the media through which the sound travels. The propagation speeds of the media are unknown; therefore, no conclusions can be made about the incident angle. The density of a medium *helps* to determine the propagation speed, but knowledge of only the density, without knowledge of the tissues' stifnesses, is insufficient to deduce which of the two media has a higher propagation speed.

246. **False.** The acoustic impedance of gel is lower than that of both the active element and the matching layer. The coupling gel acts to decrease the amount of reflection at the skin/matching layer boundary. Therefore, the gel must have an impedance between those of the matching layer and the skin. To maximize the transmission of the acoustic pulse into the body, the impedances of the active element, matching layer, gel, and skin must be in decreasing order: the impedance of the piezoelectric crystal is the highest, followed by the matching layer's impedance, the gel's impedance, and the impedance of the skin.

247. True or False? The acoustic impedance of the skin is greater than the acoustic impedances of both the matching layer and the piezoelectric element.

248. In soft tissue, sound travels to a reflector and back to the transducer in 39 μseconds. How deep is the reflector?
   a) 2 cm
   b) 6 cm
   c) 3 cm
   d) it cannot be determined

249. An ultrasonic pulse is traveling in soft tissue. Which of the following is most important in determining the frequency of the sound contained in the pulse?
   a) the propagation speed of the ultrasound transducer's matching layer
   b) the thickness of the transducer's backing material
   c) the impedance of the transducer's matching layer
   d) the propagation speed of the transducer's active element

250. What is the best estimate of the propagation speed of the ferroelectric element of a transducer used in a typical diagnostic imaging laboratory?
   a) 1.2 mm/μsec
   b) 4.0 m/sec
   c) 4.0 mm/μsec
   d) 10 hm/sec

With regard to the backing material of a pulsed ultrasound transducer, are the following four statements true or false?

251. The backing material helps to decrease the duty cycle at a particular pulse repetition frequency (PRF).

252. The backing material decreases the quality factor.

253. The spatial pulse length is decreased with the application of backing material.

254. The pulse duration is increased as a result of the presence of backing material.

247. **False.** The acoustic impedance of the skin is lower than that of both the matching layer and the piezoelectric crystal.

248. **C.** Every 13 μseconds, sound makes a round trip to a depth of 1 centimeter. Thus, in 39 μseconds, sound travels 3 cm round trip, or a total distance of 6 cm.

249. **D.** The two entities that primarily determine the frequency of sound in a pulse are: 1) the thickness of the ultrasound crystal and 2) the propagation speed of the crystal. The speed and impedance of the matching layer and the thickness of the backing material are not primary determinants of the frequency of an ultrasound pulse.

250. **C.** The propagation speed of the active element in a typical diagnostic imaging transducer is approximately three to five times greater than the speed of ultrasound through soft tissue. This range is approximately 4 to 8 km/sec. Other ways of stating this are 4 to 8 mm/μsec, or 4,000 to 8,000 m/sec. Choices A, B, and D are all too low to be the propagation speed of sound in an imaging crystal.

251. **True.** The backing material of an imaging transducer tends to shorten the duration of the pulse. This acts to decrease the percentage of the time that the system is "talking" and therefore decreases the duty factor.

252. **True.** The backing material that is attached to the back of the piezoelectric crystal will decrease the transducer's Q-factor.

253. **True.** The backing material tends to diminish the "ringing" of the transducer's active element after it is excited by the electrical signal from the pulser. This damping shortens the length of the pulse. The action of the damping material is similar to the action of a cymbal player in a band. When a short tone is desired, the musician pinches the cymbals between his arms and ribs after striking them together. This stops the cymbals from vibrating and shortens the spatial pulse length. Note that the terms *backing material* and *damping material* are synonymous.

254. **False.** The pulse duration, or the interval of time that the crystal is producing sound, is diminished when backing material is applied. When a ferroelectric crystal does not have backing material, it can vibrate freely for a long time. The backing material inhibits the crystal's oscillations and shortens the pulse duration.

255. True or False? The damping material helps to decrease the pulse repetition period achieved at a given imaging depth.

256. Which component of an ultrasound transducer is made from a slab of epoxy imbedded with tungsten particles?
    a) the matching layer   b) the piezoelectric crystal
    c) the damping material   d) the computer chips

257. Which component of an ultrasound system is made of a slab of lead zirconate titanate (PZT)?
    a) transducer's matching layer   b) transducer's active element
    c) transducer's damping material   d) scan converter's computer chips

258. True or False? The purpose of the backing material of an ultrasound transducer is to shorten the pulses, thereby creating images with better image quality.

259. You are asked to fabricate a pulsed ultrasound transducer that produces sound waves with the highest possible frequency. Which piezoelectric crystal would you select?
    a) 6 mm thick, 4 cm diameter, 4.0 mm/us propagation speed
    b) 8 mm thick, 2 cm diameter, 6.0 mm/us propagation speed
    c) 4 mm thick, 9 cm diameter, 5.0 mm/us propagation speed
    d) 2 mm thick, 6 cm diameter, 6.0 mm/us propagation speed

260. True or False? In general, ultrasound imaging transducers have a lower quality factor and a higher bandwidth than do therapeutic ultrasound transducers.

261. Assume that the frequency of sound with the greatest power emitted by a transducer is 5 MHz. However, the transducer produces acoustic energy with frequencies as low as 3.5 MHz and as high as 6.5 MHz. What is the bandwidth of the transducer?
    a) 6.5 MHz   b) 5.0 MHz
    c) 3.5 MHz   d) 3.0 MHz

255. **False.** The pulse repetition period is the time interval between the beginning of one pulse and the beginning of the subsequent pulse. The pulse repetition period is determined primarily by the maximum imaging depth desired by the sonographer and will not change unless the sonographer alters the imaging depth. It is not affected by the presence of backing material.

256. **C.** The backing material of a transducer assembly is often fabricated of tungsten-embedded epoxy.

257. **B.** Lead zirconate titanate, abbreviated PZT, is a man-made piezoelectric material commonly used as the active element of ultrasound transducers.

258. **True.** Diagnostic imaging transducers are especially effective in creating excellent images when their pulses are short. The backing material reduces the ringing of the active element and shortens the pulse duration.

259. **D.** The main frequency (also called resonant, natural, or center frequency) emitted by a piezoelectric crystal operating in the pulsed mode is principally determined by the thickness and propagation speed of the crystal. The frequency is greater when the crystal is thin and its propagation speed is high. The active element listed in selection D is therefore the proper choice. The diameter of the active element does not affect the frequency of the pulse.

260. **True.** Diagnostic imaging transducers have a lower quality factor than therapeutic ultrasound transducers. In diagnostic imaging, the pulses produced by the transducer must have short durations and spatial pulse lengths. Without this, the quality of the image would be very poor. In order to produce a short pulse, backing material is attached to the piezoelectric crystal. The backing material acts to shorten the pulse but concurrently increases the bandwidth and decreases the Q factor.

261. **D.** The bandwidth of a pulse is defined as the range of frequencies that are present within the pulse. Bandwidth is calculated by subtracting the lowest frequency from the highest frequency in the acoustic signal (6.5 - 3.5 = 3.0 MHz). The bandwidth is 3.0 MHz. The natural or center frequency of this transducer is 5 MHz.

262. Assume that the frequency of sound with the greatest power emitted by a transducer is 5 MHz. However, the transducer produces acoustic energy with frequencies as low as 3.5 MHz and as high as 6.5 MHz. What is the quality factor of the transducer?
   a) 5   b) 5/3.5
   c) 5/6.5   d) 5/3

263. Damping material is secured to piezoelectric material during the fabrication of an ultrasonic imaging transducer. What are the consequences of this attachment?
   a) bandwidth decreases and quality factor decreases
   b) bandwidth increases and quality factor decreases
   c) bandwidth decreases and quality factor increases
   d) bandwidth increases and quality factor increases

264. The impedance of the matching layer of an ultrasound transducer is 2.6 MRayls, and the impedance of the piezoelectric crystal is 3.4 MRayls. If this is assumed to be a good imaging system, what is the best estimate for the impedance of the skin?
   a) 1.5 MRayls   b) 3.8 MRayls
   c) 3.4 MRayls   d) 2.8 MRayls

265. True or False? The piezoelectric crystal of a transducer typically has an impedance higher than that of skin.

266. The action that will result when a piezoelectric crystal loses its special properties is _____ .
   a) breaking it in pieces   b) exposing it to high temperatures
   c) exposing it to electrical current d) exposing it to low pressures

267. Which temperature is closest to the Curie temperature of the common piezoelectric material known as PZT?
   a) 129^0 Fahrenheit   b) 300^0 Kelvin
   c) 300^0 centigrade   d) 300^0 Fahrenheit

268 Which properties of the piezoelectric crystal of a continuous wave transducer result in the highest emitted acoustic wave frequency?
   a) thin, high propagation speed   b) thick, slow propagation speed
   c) thin, slow propagation speed   d) none of the above

262. **D.** The quality factor of a transducer is defined as the main frequency divided by its bandwidth. In this example, the main frequency is 5 MHz and the bandwidth is 6.5 - 3.5 = 3.0 MHz (highest frequency minus lowest frequency equals the bandwidth). The Q factor is therefore 5/3.

263. **B.** The damping material of an imaging transducer increases the range of frequencies contained in the acoustic pulse (the bandwidth) and decreases the quality factor.

264. **A.** The impedance of a transducer's matching layer should be between the impedance of skin and the impedance of the piezoelectric crystal.

265. **True.** Impedance is calculated by multiplying the density of a material by the propagation speed of the material. The density of lead zirconate titanate (a common piezoelectric material) is greater than the density of skin. The propagation speeds of piezoelectric crystals are usually three to four times greater than those of the skin. Therefore, the impedance of piezoelectric crystals typically exceeds that of skin.

266. **B.** When a piezoelectric crystal is exposed to high temperatures, it will become *depolarized* and permanently lose its piezoelectric properties. This temperature is known as the Curie temperature.

267. **C.** The Curie temperature of PZT is in the range of 300° to 400° centigrade. This is approximately 600° to 700° Fahrenheit. An ultrasound transducer should never be sterilized in an autoclave. This could result in serious damage to the transducer.

268. **D.** The frequency of sound emitted by a continuous wave transducer is determined only by the frequency of the electrical signal that excites the piezoelectric crystal. The crystal's thickness and propagation speed do not affect the frequency of sound from a continuous wave transducer.

269 The region from the transducer to the location of ultrasound beam's smallest cross-sectional area is called the _____ .
    a) focus
    b) half-value thickness
    c) near zone
    d) Fraunhofer zone

270. The area that starts at the beam's smallest diameter and extends deeper is:
    a) the distant zone
    b) the Fresnel zone
    c) the Fraunhofer zone
    d) the depth of penetration

271. Match the following terms or zones that describe the same regions of an ultrasound beam.
    1) far zone
    2) near zone
    3) Fresnel zone
    4) focal zone
    5) Fraunhofer zone
    6) lateral zone
    a) 1 and 2, 3 and 4
    b) 1 and 4, 2 and 3
    c) 4 and 6, 5 and 1
    d) 2 and 3, 5 and 1

272. What is the point or location where a beam reaches its smallest dimension?
    a) near zone
    b) focus
    c) penetration depth
    d) focal zone

A PZT crystal in the shape of a disc produces a continuous ultrasound wave. The beam is unfocused. Based on this, are the following six statements true or false?

273. The near zone is the only region wherein the diameter of the sound beam is smaller than the diameter of the transducer.

274. The far zone is the only section wherein the diameter of the sound beam exceeds the diameter of the transducer.

275. If the narrowest diameter of a sound beam is located at a distance of 8 cm from the transducer face, then the PZT crystal has a diameter of 16 cm.

276. If the diameter of an acoustic beam is 8 mm at a depth equal to twice the near zone length, then the piezoelectric crystal producing the wave has a diameter of 16 mm.

277. If the diameter of the acoustic beam produced by a crystal is 8 mm at a depth of twice the near zone length, then the piezoelectric crystal has a diameter of 8 mm.

278. The near zone is the only region wherein the diameter of the sound beam decreases as depth increases.

269. C. The region or space from the transducer face to the narrowest portion of an ultrasound beam is defined as the near zone.

270. C. The region that extends deeper from an ultrasound beam's narrowest diameter is called the Fraunhofer zone.

271. D. Another name for the Fresnel zone is the near zone. The Fraunhofer zone can also be called the far zone. (Hint: Fresnel is the short name -- it is the near zone. Fraunhofer is the long name -- it is the far zone.)

272. B. The actual *location* where an ultrasound beam reaches its minimum diameter and cross-sectional area is called the focus.

273. False. The ultrasound beam created by a disc-shaped crystal operating in the continuous mode is smaller than the crystal in both the near zone and the initial part of the far zone -- up to two near zone lengths from the transducer.

274. True. The beam diameter never exceeds the transducer diameter in the near zone. The beam diameter exceeds the transducer diameter only in the far zone.

275. False. The depth of the focus is not determined solely by the diameter of the piezoelectric crystal producing it.

276. False. When a continuous wave is produced by a disc-shaped crystal, the beam's diameter equals the crystal's diameter at a depth of twice the near zone length.

277. True. When a continuous wave is produced by a disc-shaped crystal, the beam's diameter equals the crystal's diameter at a depth that is twice the near zone length.

278. True. In the near zone, the sound beam actually tapers down until the focus is reached. Once at the focus, the end of the near zone is reached and the beam diameter increases from that point outward.

A PZT crystal in the shape of a disc produces a continuous ultrasound wave. The beam is unfocused. Based on this, are the following three statements true or false?

279. The higher the frequency of the acoustic wave, the shorter the length of the near zone.

280. The greater the diameter of a transducer's piezoelectric crystal, the longer the length of the near zone.

281. The thicker a continuous wave transducer's piezoelectric crystal, the longer the near zone length.

282. When a disc-shaped piezoelectric crystal produces a continuous acoustic wave, which design will produce a beam with the most shallow focus?
    a) large diameter, low frequency   b) large diameter, high frequency
    c) small diameter, low frequency   d) small diameter, high frequency

283. As a sound wave travels deep into the far zone, it tends to diverge or spread out.' Which of the following will result in a minimum beam divergence deep in the far zone?
    a) small diameter   b) high frequency
    c) large diameter   d) low frequency

For the following three statements, indicate the appropriate transducer type. (More than one answer may be correct.)
    a) linear switched array transducer
    b) linear phased array transducer
    c) mechanical transducer
    d) annular phased array transducer
    e) convex or curvilinear array transducer

284. With these transducers, the ultrasound beam is steered by a motor, a reflecting mirror, or a similar device.

285. These transducers focus the ultrasound beam electronically.

286. With these transducers, the ultrasound beam is focused with an acoustic lens, a mirror, or by milling the piezoelectric crystal in a curved shape.

279. **False.** The focus of a continuous acoustic beam produced by a disc-shaped crystal gets deeper as the frequency of the wave increases. When a low frequency wave is produced, the focus is closer to the transducer face.

280. **True.** Large-diameter crystals tend to produce continuous wave sound beams with a deeper focus. The beam's near zone length is longer.

281. **False.** Increasing the thickness of a piezoelectric crystal does not alter the length of the near zone of a continuous wave ultrasound beam.

282. **C.** A continuous wave transducer producing the shortest near zone (the most shallow focus) will have a small diameter and a low emitted frequency.

283. **C.** Large-diameter crystals tend to produce sound waves that diverge minimally in the deep far zone. In contrast, waves from small diameter crystals diverge remarkably in the deep far zone.

284. **C and D.** Steering an ultrasound beam is achieved by using mechanical devices, such as motors or mirrors, in both mechanical and annular phased array transducers.

285. **B, D, and E.** Electronic focusing techniques are employed in linear phased array and annular phased array transducers. Focusing an acoustic beam produces images with greater detail. Some convex array transducers focus the beam electronically, whereas others do not.

286. **A, C, and E.** Acoustic beams can be focused with techniques similar to those that focus light waves, including using an acoustic lens, a focusing mirror, or a piezoelectric crystal that is manufactured in a curved shape. Linear sequential, mechanical, and some curvilinear array transducers use these techniques to focus acoustic beams.

For the following seven statements, indicate which of the transducer types are
   appropriate. (More than one answer may be correct.)
   a) linear switched array transducer
   b) linear phased array transducer
   c) mechanical transducer
   d) annular phased array transducer
   e) convex or curvilinear array transducer
   f) vector array transducer

287. With these transducers, a rectangular image shape is routinely produced.

288. With these transducers, a wedge-shaped or "slice of pie" image is
   produced. At its origin, the wedge is very narrow and tapers to a point.

289. A special wedge shape is produced when this transducer is used. It looks
   like a wedge or a "slice of pie," but it doesn't originate at a point. The
   sector has a blunted, curved shape. The beginning of the image may be a
   few centimeters wide.

290. With these transducers, a trapezoidal-shaped image is produced. The
   image is flat in the region adjacent to the transducer and becomes
   progressively wider at increasing depths.

291. Ultrasound systems using these transducers do not actually steer the
   ultrasound beams.

292. These transducers steer the beam with a motor, mirror, or similar device
   and also focus the beam with a mirror, acoustic lens, or a curved
   piezoelectric element.

293. These transducer systems typically produce an image that, throughout its
   entire depth, is approximately as wide as the ultrasound transducer itself.

287. **A.** Linear sequential arrays commonly produce rectangular images. The width of the transducer determines the width of the ultrasound image. Rectangular images are created because an ultrasound beam is emitted from each of the crystals lined up in the array. The beam emitted from each crystal travels in a straight line and is parallel to those created by the neighboring crystals. Therefore, in producing a two-dimensional image, a linear switched or sequential transducer does not really steer the ultrasound beam, but rather uses its design to make a rectangular image.

288. **B, C, and D.** Transducers that steer beams through a pathway in order to create an imaging plane usually create a pie-shaped image that starts at a point. This sector image is similar to that of spokes radiating out from the hub of a bicycle wheel. The wedge shape is created by mechanical, linear phased array and annular phased array transducers.

289. **E.** Curved array transducer systems create a special type of sector image that starts at an initial width and gets progressively larger. This initial width of the sector results from the design of the transducer array. The piezoelectric crystals are arranged in a curved array that may be 2 to 4 cm in length. Therefore, the image will have a beginning width of 2 to 4 cm and then progressively widen.

290. **F.** Vector arrays create this image shape. The vector transducer is a combination of linear phased and linear switched array technologies.

291. **A.** Linear switched (or linear sequential arrays) do not steer acoustic beams in order to create a two-dimensional image. Rather, each crystal in the line produces a pulse that travels straight ahead. If the transducer is 7 cm wide, then the image will be no wider than 7 cm. The transducer architecture defines the width of the two-dimensional image.

292. **C.** Only a mechanical scanner has the characteristics described above. Beam steering is achieved by using a mechanical technique (i.e., a motor or a moving reflecting mirror). Focusing is also accomplished with a traditional method such as a lens, mirror, or curvature of the PZT.

293. **A.** The width of a linear sequential array transducer determines the width of the 2-D image that it creates. Since the beam is not redirected as it leaves the piezoelectric crystal, the width of the image remains equal to that of the transducer throughout the entire imaging depth.

294. When viewing images produced by the following transducer systems, which one can be distinguished from the others?
    a) linear phased array   b) linear switched array
    c) annular phased array   d) mechanical transducer

295. True or False? All phased array transducer systems direct ultrasound pulses in many directions to create a two-dimensional image.

296. What is true of the electrical pulses that excite the active elements of an annular phased array transducer?
    a) they arrive at each piezoelectric crystal at exactly the same time
    b) they vary in amplitude based on the direction of the steering and focusing
    c) they excite the crystals at different times, tiny fractions of a second apart
    d) they arrive at each piezoelectric crystal at different times, up to 0.5 seconds apart

297. The pattern of electrical signals exciting the piezoelectric crystals of a linear phased array transducer _____.
    a) typically changes from one acoustic pulse to the next
    b) changes every fourth or fifth acoustic pulse
    c) changes only when the maximum imaging depth changes
    d) none of the above

298. How are the piezoelectric crystals excited when an ultrasound system with a sequential linear array transducer is used?
    a) singly and in order: the first crystal, the second, then the third, etc.
    b) in pairs: the first two crystals, the 3rd & 4th together, the 5th & 6th
    c) in a specific order
    d) in a random sequence

299. How are the piezoelectric crystals of a linear phased array transducer fired?
    a) in a single specific pattern   b) in order, from top to bottom
    c) at exactly the same time   d) at almost exactly the same time

294. **B.** The linear switched array (also called linear sequential array) transducer system produces an image of rectangular shape. The other three transducers generate similar sector-shaped images that are essentially indistinguishable from each other.

295. **True.** In order to create a two-dimensional image, the ultrasound beam of both annular, linear, and curvilinear phased array transducers must be steered through a path that defines the imaging plane. The linear and curvilinear phased array systems steer the beam electronically, whereas the annular phased array systems steer the beam mechanically.

296. **C.** Annular array technology produces a focused ultrasound beam by firing the piezoelectric crystals in a particular sequence. The electrical pulses that excite the ringed elements of the array arrive at slightly different times, separated by only very small fractions of a second.

297. **A.** The pattern of the electrical signals that excite the piezoelectric crystals of a linear phased array transducer changes from one pulse to the next. The electrical patterns determine the direction and focusing of each acoustic pulse. Each sound wave is directed in a slightly different pattern, resulting in a two-dimensional imaging plane. Thus, the electrical signals vary for each acoustic pulse produced.

298. **C.** Sequential linear array transducers fire their piezoelectric crystals in a specific succession or progression. The order of firing is determined by the manufacturer and is chosen to produce the highest quality image. There is, however, no specific order that all linear arrays use. The firing pattern is best described as "in a specific order."

299. **D.** To steer and focus the ultrasound beam produced by a linear phased array transducer, the separate piezoelectric crystals are excited by electronic pulses at almost the same time. Only fractions of thousandths of seconds separate the electronic pulses that strike the crystal elements in the transducer assembly.

With regard to a mechanical transducer, are the following four statements true or false?

300. The ultrasound beam is directed in many different directions to create an imaging plane.

301. The ultrasound beam is not electronically focused.

302. There is communication between the pulser of the ultrasound system and the piezoelectric crystal of the transducer.

303. Foci exist at multiple depths as an ultrasound beam propagates through the body.

304. What helps to determine the frequency of the sound produced by the transducer of a continuous wave ultrasound system?
    a) piezoelectric crystal diameter   b) piezoelectric crystal thickness
    c) damping material density   d) ultrasound system electronics

305. In pulsed wave ultrasonic imaging, what helps to establish the primary frequency of the acoustic energy discharged by the transducer?
    a) piezoelectric crystal diameter   b) piezoelectric crystal thickness
    c) damping material density   d) ultrasound system electronics

306. What helps to determine the frequency of the acoustic wave produced by a continuous wave transducer?
    a) the thickness of the transducer's active element
    b) the frequency of the electrical signal exciting the piezoelectric crystal
    c) the impedance of the medium and the piezoelectric crystal thickness
    d) the piezoelectric crystal's propagation speed and the matching layer's impedance

307. True or False? Both linear and phased array technologies have the ability to variably focus an ultrasound beam to different depths.

308. Annular phased array transducers have active elements that are in the shape of _____.
    a) rectangles   b) squares
    c) wedges   d) circles

300. **True.** With a mechanical transducer, the active element is aimed in a variety of directions so that the pulses can create a two-dimensional picture. Without this beam steering, the pulses would travel in the same direction and produce only a single B-mode line on the display.

301. **True.** With standard mechanical transducers, focusing is not achieved electronically. The sound beam is focused by using an acoustic lens, a mirror, or by building the PZT material in a curved shape (internal).

302. **True.** It is essential for the transducer's PZT to be directly connected to the pulser electronics of the US system. This wire provides a route for the electrical signal to travel from the pulser to the PZT in order to produce an acoustic pulse. In addition, the crystal must send information back to the US system regarding the reception of reflected echoes. This data is sent to the receiver, rather than to the pulser, to create an image.

303. **False.** A standard mechanical transducer cannot focus at multiple depths, nor can it provide for variable or user-selectable focal depths. Mechanical transducers support only a single, unalterable focal depth.

304. **D.** A continuous wave transducer produces an acoustic wave with a frequency equal to that of the electrical signal that excites the crystal. When the pulser's electrical signal has a frequency of 6 MHz, then the emitted acoustic wave is also 6 MHz. In a continuous wave US system, the electronics determine the frequency of the acoustic wave.

305. **B.** The frequency of the acoustic wave produced by a standard pulsed wave imaging instrument is partly determined by the thickness of the piezoelectric element. Just as different tones are produced when a musician plays different bars on a xylophone, piezoelectric crystals of various thicknesses produce acoustic pulses of different frequencies.

306. **B.** In the case of continuous wave ultrasound, the frequency of the acoustic signal produced by the piezoelectric crystal will be equal to the frequency of the electrical signal driving the crystal.

307. **True.** With the phased delivery of electrical pulses to the crystals, the ultrasound beam may be focused at a variety of different depths.

308. **D.** Annular array transducers have piezoelectric crystals that are donut- or ring-shaped. They are arranged as a collection of concentric rings.

309. The minimum number of active elements in a mechanical transducer is:
a) 0   b) 1
c) 2   d) none of the above

310. Two ultrasound systems produce acoustic pulses. One pulse is 0.4 μsec in duration, whereas the other is 0.2 μsec in duration. Which pulse will most likely provide the best azimuthal resolution?
a) 0.4 μsec pulse   b) 0.2 μsec pulse
c) they are the same   d) it cannot be determined

311. You are purchasing a diagnostic ultrasound system. System X has a longitudinal resolution of 0.7 mm, and System D has a longitudinal resolution of 0.4 mm. Based on this information, which system will produce the better quality picture?
a) System X   b) System D
c) they have the same quality d) it cannot be determined

312 Two imaging systems produce acoustic pulses: the duration of one pulse is 0.4 μsec, and the duration of the other is 0.2 μsec. Which is most likely to provide the best temporal resolution?
a) 0.4 μsec system   b) 0.2 μsec system
c) they are the same   d) it cannot be determined

313. Two ultrasound systems produce pulses. One pulse is 0.4 μsec in duration, whereas the other is 0.2 μsec in duration. Which pulse is most likely to provide the best radial resolution?
a) 0.4 μsec system   b) 0.2 μsec system
c) they are the same   d) it cannot be determined

314. The radial resolution of an ultrasonic imaging system is reported to be 0.85 mm at the beam's focus. What is the closest estimate of the system's radial resolution at a location 5 cm deeper than the focus?
a) less than 0.85 mm   b) equal to 0.85 mm
c) greater than 0.85 mm

315. Two ultrasound systems produce acoustic pulses. A pulse from System 1 has a wavelength of 0.5 mm, 4 cycles per pulse, and a pulse repetition period of 1.2 msec. The pulse from System 2 has a wavelength of 1.0 mm, 2 cycles per pulse, and a pulse repetition period of 1.8 msec. Which ultrasound system will have a lower numerical value of range resolution?
a) System 1   b) System 2
c) both are the same   d) it cannot be determined

309. **B.** A mechanical transducer may be constructed with a minimum of a single piezoelectric crystal. An imaging plane may be created by steering the single crystal with a motor.

310. **D.** The duration of a pulse does not determine the azimuthal resolution. Pulse duration affects longitudinal resolution. Therefore, it is not possible to answer this question with the information provided.

311. **B.** Resolution is reported in units of distance, such as cm or mm. Higher image quality is achieved by systems with lower numerical values for resolution. System D has a lower numerical value for longitudinal resolution and will produce more detailed, higher quality pictures.

312. **D.** Temporal resolution is not affected by the pulse duration. Temporal resolution is determined by the number of frames, images, or pictures that are produced each second. Therefore, there is insufficient information to answer this question.

313. **B.** Radial resolution is determined by the pulse duration or the spatial pulse length. The shorter the time span that a pulse exists (or the shorter the length of the pulse), the better the resolution of the ultrasound system. Therefore, the device producing the shorter pulse, 0.2 μsec, has the best radial resolution.

314. **B.** Radial resolution is determined by the spatial pulse length or the pulse duration. These variables remain constant, regardless of the depth of the pulse. Therefore, the radial resolution is the same at all imaging depths.

315. **C.** Range resolution is equal to one-half of the spatial pulse length. Spatial pulse length is equal to the number of cycles in the pulse multiplied by the wavelength. The pulse length of system 1 is 4 x 0.5 mm, or 2.0 mm. The pulse length of system 2 pulse length is 2 x 1.0 mm, or 2.0 mm. Because the systems have identical pulse lengths, their range resolutions are also identical.

316. An ultrasonic pulse has a pulse repetition period of 1.2 msec, a spatial pulse length of 2.0 mm, and a wavelength of 0.4 mm. What is the radial resolution of the system?

   a) 2.0 mm   b) 1.0 mm

   c) 0.4 mm   d) 1.8 mm

317. A sonographer is performing a study on a patient and desires superior depth resolution. Which of the following changes would create such a system?

   a) higher frequency   b) shorter wavelength

   c) fewer cycles per pulse   d) all of the above

318. True or False? The lower the numerical value of the longitudinal resolution, the worse the picture produced by an ultrasound system.

319. True or False? The more cycles there are in a pulse, the greater the detail that will be visualized in the ultrasound scan.

320. True or False? The pulse duration does not profoundly influence the lateral resolution.

321. True or False? The higher the frequency of the cycles within a pulse, the lower the value of the axial resolution.

322. True or False? The shorter the pulse length, the better the picture.

316. **B.** The radial resolution of an ultrasound system is equal to half of the spatial pulse length produced by the system. In this example, the spatial pulse length is equal to 2.0 mm, and the radial resolution is 1.0 mm.

317. **D.** For higher quality pictures, the numerical value for the depth resolution is small. Systems that produce shorter pulses create better pictures and have lower values for depth resolution. Higher frequency, shorter wavelength, and fewer cycles per pulse all shorten the spatial pulse length, decrease the numerical depth resolution, and make better images.

318. **False.** Higher quality images are associated with lower values of longitudinal resolution. The numerical value indicates how close together two structures can be and still produce two distinct images on the ultrasound display. Therefore, lower numbers identify ultrasound systems that display images with fine detail.

319. **False.** A pulse with many cycles is long. Systems with long pulses cannot produce images with fine detail. When there are many cycles in a pulse, less detail is presented on the scans.

320. **True.** In general, the primary determinant of lateral resolution is the diameter of an ultrasound pulse. The length of the pulse does not dramatically alter the lateral resolution. It is true that higher frequency ultrasound pulses have both shorter pulse lengths and somewhat better lateral resolution. However, shorter pulse lengths do not have a profound and direct influence on lateral resolution.

321. **True.** When traveling through a particular medium, higher frequencies produce cycles with shorter wavelengths. Shorter wavelengths will shorten the spatial pulse length. Since the axial resolution is equal to one-half of the spatial pulse length, the numerical value of the axial resolution will also be reduced.

322. **True.** This is the fundamental concept of longitudinal resolution. Shorter pulses produce better pictures. In fact, one of the most important design criteria for ultrasound systems and transducers is to minimize the pulse duration and spatial pulse length.

323. True or False? One way that a sonographer can alter the axial resolution achieved during an exam is to adjust the maximum imaging depth.

324. True or False? The shorter the pulse duration, the better the picture.

325. True or False? The length of a pulse does not directly influence the temporal resolution.

326. True or False? With a specific ultrasound system and transducer, the system's range resolution is invariant: therefore, the sonographer can do nothing to improve it.

327. The lateral resolution of an ultrasound system is primarily determined by the _____ .
    a) width of the sound pulse   b) length of the ultrasound pulse
    c) duration of the sound pulse   d) none of the above

328. The focus of an ultrasound beam is the location where the _____ .
    a) beam is the broadest   b) optimum transverse resolution is found
    c) frequency is the highest   d) finest depth resolution is obtained

329. Two ultrasound systems have near zone lengths of 8 cm. At the focus, System G has a lateral resolution of 3.0 mm, whereas System P has a lateral resolution that measures 5.0 mm. Which system is most likely to produce higher quality pictures at their foci?
    a) System P
    b) System G
    c) both will produce similar quality pictures

330. True or False? In comparison with other locations along the length of an ultrasound beam, focusing is generally ineffective in the far zone.

323. **False.** The only way to alter the axial resolution of an ultrasound system is to alter the pulse duration. This is not possible when using any of the controls on the ultrasound system. A sonographer can change a system's axial resolution only by selecting a different transducer that produces pulses with a different duration.

324. **True.** This restates one of the fundamental principles of ultrasonic imaging. The shorter the pulse duration, the higher the quality of the images.

325. **True.** Temporal resolution is not affected by the length of the pulses produced by the system. The factors that do influence the temporal resolution are maximum imaging depth, sector angle, line density, and the number of foci per scan line.

326. **True.** The range resolution of an ultrasound system is determined only by the components of the system. While using a particular ultrasound system, the operator cannot change range resolution.

327. **A.** The lateral resolution achieved by an imaging system is approximately equal to the width of the ultrasound pulse.

328. **B.** The focus of an ultrasound beam is the location where the beam is most narrow. The narrowest portion of the beam provides the optimal transverse resolution.

329. **B.** Because the beam is narrower in System G than in System P at the end of the near field (the location of the foci), System G will produce higher quality images at that depth.

330. **True.** Focusing is a technique that narrows the width of an ultrasound beam in order to improve image quality. When a beam undergoes focusing, it narrows in the region close to the end of the near zone and the beginning of the far zone.

331. Two ultrasound systems have near zone lengths of 8 cm. At the focus, System S has a lateral resolution of 3.0 mm whereas System C has a lateral resolution that measures 5.0 mm. Which system is most likely to appropriately display two small body structures, lying one in front of the other, at depths of 8.6 and 9.0 mm?
    a) System C   b) System S
    c) both have comparable quality   d) it cannot be determined

332. Two ultrasound systems are being evaluated. Both have near zone lengths of 8 cm. At their foci, System Q has a lateral resolution of 3.0 mm, and System H has a lateral resolution of 5.0 mm. Which system will correctly display two small structures that lie in the body at a depth of 8 cm? (The objects are side-by-side and are 0.4 cm apart.)
    a) System H   b) System Q
    c) both will produce similar quality pictures

333. The lateral resolution of an US system is 4 mm. Two structures are separated by 3 mm and lie side-by-side in relation to the long axis of the acoustic beam. What will most likely appear on the display of the system?
    a) two weak echoes, 4 mm apart   b) two bright echoes, 3 mm apart
    c) one echo   d) two bright echoes, 7 mm apart

334. True or False? When using standard ultrasonic imaging instrumentation, the lateral resolution has a higher value than does the axial resolution.

335. True or False? When using an instrument typical of today's imaging devices, a higher frequency transducer is likely to mildly improve the system's transverse resolution.

336. True or False? When using an instrument typical of today's diagnostic imaging devices, a higher frequency transducer is likely to improve the system's range resolution.

337. What will lower the value of an ultrasound system's lateral resolution?
    a) decrease the number of cycles in the pulse
    b) increase the effective damping material
    c) increase the pulse repetition period
    d) send the pulse through an acoustic lens

331. **D.** There is insufficient information to answer this question. The information provided gives insight into the lateral resolution of the systems. However, the question asks about the ability to distinguish two structures that lie one in front of the other. This depends upon longitudinal resolution.

332. **B.** At a depth of 8 cm, System Q will produce a more accurate picture of two structures that are 0.4 cm apart . A system's lateral resolution is approximated by the beam diameter at that depth. System Q's beam diameter is 3.0 mm at its focus and will display this pair of reflectors accurately. System H's beam, with a diameter of 5 mm, is too wide to distinguish this pair of reflectors as distinct and separate.

333. **C.** When two structures are closer together than the resolution of an ultrasound system, the images of the structures will blur together. Since these structures are 3 mm apart, while the lateral resolution is 4 mm, only a single echo representing both reflectors will appear on the image.

334. **True.** Most imaging systems have better axial resolution than lateral resolution. Therefore, the lateral resolution will have a higher numerical value than the axial resolution. This occurs because a sound pulse is typically wider than half of its length.

335. **True.** With all other variables being equal, pulses made with higher frequency sound are somewhat narrower than low-frequency pulses. Narrow beams improve the transverse resolution of the imaging system.

336. **True.** Higher frequency ultrasound pulses typically have shorter pulse lengths. Short pulse lengths correspond to images with good range resolution.

337. **D.** To lower the numerical value of lateral resolution of an ultrasound system, the diameter of the beam must be reduced. A commonly used technique to decrease the diameter of the beam is to focus the beam with an acoustic lens.

338. Which of the following will improve a system's temporal resolution?
    a) increased sector angle   b) increased line density
    c) increased PRF   d) increased frequency

339. In which region of a sound beam is focusing is most effective?
    a) the very shallow near zone   b) the end of the near zone
    c) very deep in the far zone   d) throughout its entire length

340. The diameter of a disc-shaped, unfocused piezoelectric crystal is 1 cm. What is the best estimate for the minimum lateral resolution of the ultrasound system?
    a) 1 mm   b) 5 mm
    c) 1 cm   d) 5 cm

341. The diameter of a unfocused, disc-shaped piezoelectric crystal is 1.2 cm. The near zone length is 8 cm. What is the best estimate for the lateral resolution at a depth of 16 cm?
    a) 0.6 cm   b) 1.2 cm
    c) 8 cm   d) 16 cm

342. Two ultrasound systems are identical except for the diameter of the transducer's piezoelectric crystal. Which system has the farthest focus?
    a) the system with the smaller diameter crystal
    b) the system with the larger diameter crystal
    c) their foci will be at the same depth

343. Two ultrasound systems are identical except for the frequency of the emitted pulse. Which system will have the farthest focus?
    a) the lower frequency system
    b) the higher frequency system
    c) both foci will be at the same depth

344. Two ultrasound systems are identical except for the pulse repetition period of the emitted pulse. Which system will have the farthest focus?
    a) the system with the lower pulse repetition period
    b) the system with the higher pulse repetition period
    c) their foci will be at the same depth

338. **C.** Improved temporal resolution is achieved by increasing the number of images produced each second. If the number of pulses emitted per second is increased (choice C), then the system has the capacity to increase the frame rate. Increased PRF is obtained by decreasing the maximum imaging depth of the exam. Selections A, B, and D would all decrease the temporal resolution.

339. **B.** The region of a beam that is most affected by focusing is close to the end of the near zone and the beginning of the far zone. This region is called the focal zone.

340. **B.** Generally, unfocused sound beams produced by a disc-shaped crystal can be viewed as having an hourglass shape where the beam tapers to a minimum diameter at the focus. At the focus, the beam diameter is approximately one-half of the diameter of the piezoelectric crystal. In this example, the minimum beam diameter would be approximately 5 mm, and this represents the lateral resolution of the system.

341. **B.** Generally, unfocused sound beams produced by a disc-shaped crystal have a profile where the beam tapers to a minimum diameter at the focus and then starts to diverge. At a distance from the transducer of twice the near zone length, the beam diameter is approximately equal to the piezoelectric crystal diameter. The beam diameter at a depth of 16 cm (which is twice the near zone length) is equal to the crystal diameter, 1.2 cm. This equals the lateral resolution at that depth.

342. **B.** Large-diameter crystals tend to produce ultrasound beams with more remote foci. The larger the crystal diameter, the farther the focus. Smaller diameter crystals produce beams with shallower foci.

343. **B.** High-frequency piezoelectric crystals tend to produce ultrasound beams with more remote foci. The higher the acoustic frequency, the farther the focus. Lower frequency crystals produce beams with shallow foci.

344. **C.** Changes in the pulse repetition period do not affect the depth of the focus. The frequency of the sound and the diameter of the piezoelectric crystal help to determine the focal depth.

345. True or False? An ultrasound system with a longer pulse duration will generally have better temporal resolution.

346. True or False? An ultrasound system with a lower pulse repetition period will generally have better temporal resolution than a system with a higher pulse repetition period.

347. What is the primary advantage of multiple focal zones along each scan line of a two dimensional image?
    a) improved temporal resolution   b) decreased temporal resolution
    c) improved lateral resolution   d) improved longitudinal resolution

348. What is the primary disadvantage of multiple focal zones along each scan line of a two dimensional image?
    a) improved temporal resolution   b) decreased temporal resolution
    c) improved lateral resolution   d) improved longitudinal resolution

349. True or False? With sector scanning, images with greater detail are created when the number of acoustic pulses per degree of sector is increased.

350. Which of the following sector imaging systems will have the best image detail if all other parameters are identical?
    a) a 90° sector with 100 pulses/image
    b) an 80° sector with 40 pulses/image
    c) a 70° sector with 84 pulses/image
    d) a 60° sector with 60 pulses/image

351. An ultrasound system with a 4.0 MHz transducer is used to image structures as deep as 15 cm. Twenty images are produced each second, each requiring 100 acoustic pulses. What is the pulse repetition frequency of the system?
    a) 1,500 Hz   b) 300 Hz
    c) 2,000 Hz   d) 4 MHz

345. **False.** Temporal resolution of a system is unaffected by the pulse duration.

346. **True.** An ultrasound system with a lower pulse repetition period has the ability to produce more pulses each second. With more pulses per second, the system may create numerous images per second. This increased frame rate results in improved temporal resolution.

347. **C.** Multiple foci along a single scan line create a composite beam that is extremely narrow. Narrow sound beams create images with superior lateral resolution at all depths. Traditional scanners use a single sound beam to create each scan line, and lateral resolution is optimal at only a single depth.

348. **B.** With multiple foci along each scan line, the US system uses many pulses to create each image. This reduces the number of frames created per second. A low frame rate results in decreased temporal resolution.

349. **True.** The more pulses per degree of sector, the better the image's spatial resolution. If the line density is low, gaps may exist between neighboring pulses. Reflectors in the gaps may not appear on the image.

350. **C.** The best image detail is provided by the ultrasound system with the highest line density. Line density is defined as the number of acoustic lines per degree of sector angle. The more lines per sector degree that a system provides, the closer the scan lines are squeezed together, and the greater the anatomical detail displayed. Choice C has 1.2 acoustic pulses per degree (84/70), choice A has a line density of 1.11 (100/90) lines/degree, choice B has 0.5 (40/80) lines/degree, and selection D has 1 (60/60) lines/degree. Choice C has the highest line density and the image with the best spatial resolution.

351. **C.** Pulse repetition frequency is defined as the number of pulses produced by an ultrasound system in one second. In this example, 100 pulses are required to construct a single image, and 20 images are displayed each second. Therefore, a total of 2,000 pulses is used each second to meet the imaging requirements of this system. This is the PRF.

352. True or False? In diagnostic imaging, it is possible to image with 200 lines/frame, 15 frames/second and a maximum imaging depth of 20 cm.

353. Which ultrasound imaging modality has the best temporal resolution?
    a) B-scanning
    b) duplex imaging
    c) color flow imaging
    d) M-mode

354. True or False? A diagnostic ultrasound system can produce 20 images per second, to a maximum depth of 20 cm, and create each image with 200 acoustic pulses.

355. Of the following, the ultrasound imaging modality with the lowest temporal resolution is _____.
    a) two dimensional, real-time imaging
    b) A-mode
    c) color flow imaging
    d) M-mode

356. Of the following, which ultrasound imaging modality has the worst temporal resolution?
    a) two dimensional real-time imaging
    b) B-scanning
    c) color flow imaging
    d) M-mode

352. **True.** To answer this question, the number of lines per frame, frame rate, and imaging depth must be multiplied together. The product of these three terms must be less than 77,000. Sound travels at exactly 154,000 cm/sec in soft tissue. Sound must travel to the reflector and return to the transducer; therefore, pulses can only travel to a depth of 77,000 cm and back in one second. 200 x 15 x 20 = 60,000 cm. An ultrasound system can perform under these settings because 60,000 is less than 77,000.

353. **D.** Temporal resolution is the ability of an ultrasound system to record the position of reflectors with regard to time. Which system tracks the position of moving structures most faithfully or accurately? With M-mode, the ultrasound pulse tracks reflector position in only a single dimension (depth only) and delivers extremely fine temporal resolution. The remaining choices all operate in the two-dimensional imaging mode and do so by compromising temporal resolution.

354. **False.** To answer this question, the number of lines per frame, frame rate, and imaging depths must be multiplied together. The product of these three terms must be less than 77,000. Sound travels at exactly 154,000 cm/sec in soft tissue. Sound travels to the reflector and returns to the transducer. Therefore, pulses can only travel to a depth of 77,000 cm and back in one second. 200 x 20 x 20 = 80,000 cm. An ultrasound system cannot perform under these settings because 80,000 is greater than 77,000.

355. **C.** Of the choices, color flow imaging is usually the modality with the lowest temporal resolution. Many pulses are required to create a single color flow image. The large number of pulses required to construct one image results in a low frame rate and reduced temporal resolution.

356. **B.** Of the choices, the modality with the lowest temporal resolution is B-scanning. The sonographer performing B-scans must move the transducer through a predetermined path to create a single two-dimensional image. This takes a relatively long time and, thus, it is impossible to perform real-time imaging using B-scanning. Temporal resolution, the ability to track or locate a moving reflector in time, is at a minimum with B-scanning.

357. Which ultrasound imaging modality places dots of dissimilar intensities on the display screen?
    a) A-mode   b) B-mode
    c) I-mode   d) M-mode

358. Which imaging technique usually displays a single jagged line of varying heights?
    a) A-mode   b) B-mode
    c) C-mode   d) I-mode

359. The scanning approach that produces lines related to the position of reflectors in the body with respect to time is called _____.
    a) A-mode   b) B-mode
    c) M-mode   d) a la mode

360. Which imaging method creates images of reflectors that are all located at the same depth in the body?
    a) A-mode   b) B-mode
    c) C-mode   d) M-mode

361. The Doppler effect is observed as a change in _____ and has units of _____.
    a) amplitude, watts   b) power, watts
    c) frequency, per second   d) wavelength, millimeters

362. In clinical imaging, which reflectors produce most relevant Doppler shifts?
    a) blood vessels   b) blood plasma
    c) platelets   d) red blood cells

363. If red blood cells are traveling toward a transducer, the frequency emitted by the transducer is _____ the frequency reflected from the red blood cells.
    a) greater than   b) equal to   c) less than

364. When red blood cells move away from a transducer, the frequency of the wave reflected from the red cells is _____ the frequency emitted by the transducer.
    a) greater than   b) less than   c) equal to

357. **B.** B-mode, or brightness mode, exhibits reflected echoes as spots of varying brightness on the display of an US system. The location of the dot is based on the time that passes between production of the pulse and the returning echo. The brightness of the dot is determined by the strength of the returning echo. The stronger the reflected pulse, the brighter the display's dot. The weaker the reflection, the darker the dot.

358. **A.** A-mode, or amplitude mode, exhibits reflected echoes as spikes or peaks of varying height on the display of an US system. A-mode displays the amplitude of the reflected echoes on the Y-axis and the depth of the reflector on the X-axis. The location of the peak is based on the time that passes between the production of the pulse and the returning echo. The height of the peak is determined by the strength of the reflected pulse. Stronger reflections create higher upward deflections. The weaker the reflection, the smaller the upward deflection.

359. **C.** The display mode that allows a sonographer to gain insight into the position of reflectors throughout time is M-mode. M-mode, or motion mode, displays the position of reflectors on the Y-axis and time on the X-axis. Therefore, M-mode is the only display process that provides reflector position throughout time.

360. **C.** C-mode, or constant depth mode processing, displays anatomical data that exists at a particular depth from the surface of the body. The reflections arising from structures at that depth are processed, while other reflections originating from other depths are discarded.

361. **C.** The Doppler effect is a change in a wave's frequency when the sound source and the receiver of the sound are in motion relative to each other. Frequency may be expressed as Hertz, Hz, cycles/sec, or per second.

362. **D.** In clinical imaging, red blood cells are the primary reflectors that produce Doppler shifts. Blood cells constantly move through the circulatory system and make up nearly 45% of the volume of the blood.

363. **C.** When RBCs approach a transducer, a positive Doppler frequency shift results. The emitted frequency is less than the reflected frequency.

364. **B.** When red blood cells are moving farther away from a transducer, the Doppler shift is negative. The reflected frequency is less than the frequency originally transmitted from the transducer.

365. What information does the Doppler shift furnish concerning the blood cells that produce it?
   a) frequency   b) speed
   c) velocity   d) density

366. A duplex ultrasound system displays _____ information.
   a) M-mode, two dimensional image, and A-mode
   b) A-mode and B-mode
   c) two dimensional image and Doppler
   d) two dimensional image and M-mode

367. In standard Doppler, what is known about the reflected frequency produced by blood cells traveling in a direction away from the transducer?
   a) it is in the audible range
   b) it is ultrasonic
   c) it is greater than the transmitted frequency
   d) it is equal to the transmitted frequency

368. What is the range of Doppler shifts commonly measured in clinical exams?
   a) -10 kHz to 1 MHz   b) -0.5 MHz to 0.5 MHz
   c) -0.01 MHz to 0.01 MHz   d) none of the above

369. A maximum Doppler shift is obtained when the angle between the direction of blood flow and the direction of the sound beam is _____.
   a) 10 degrees   b) 90 degrees
   c) 180 degrees   d) 270 degrees

370. The Doppler shift does not always provide a valid estimate of the speed of the red blood cells that produce it because the shift is related to the _____ of the angle between the direction of the beam and the direction of blood flow.
   a) sine   b) tangent
   c) cosine   d) cotangent

371. What can be said of the Doppler shift when the sound beam is normally incident to the velocity of the red blood cells?
   a) it is at a maximum   b) it is half of maximum
   c) it is absent   d) it is at minimum

365. **C.** Doppler shifts produce information regarding velocity. What is the difference between speed and velocity? Speed is purely a number such as 400 miles per hour. Velocity is both a number and a direction, such as 200 m/sec headed away from the transducer. Since Doppler shifts indicate direction as well as magnitude, they provide velocity information.

366. **C.** A duplex scanner displays both Doppler and two-dimensional image data.

367. **B.** The frequency of the sound wave reflected by the red blood cells back to the transducer is ultrasonic (having a frequency greater than 20,000 Hertz). Although the difference between the transmitted and reflected frequencies is audible, the reflected wave itself has a frequency in the ultrasonic range.

368. **C.** Doppler shifts in clinical imaging are typically in the range of -10,000 Hz to +10,000 Hz. Choice C has units of megahertz, but close examination of the values of -0.01 to +0.01 MHz reveal that they are exactly the same as -10,000 Hz to +10,000 Hz.

369. **C.** Maximum Doppler shifts occur when red blood cells travel directly toward or directly away from a transducer. An angle of 180° between the sound source and the direction of motion exists when the cells travel directly away from the transducer.

370. **C.** Included in the Doppler equation is the COSINE of the angle between the direction of motion and the direction of the sound beam.

371. **C.** The Doppler shift is related to the angle between the direction of blood flow and the direction that the acoustic wave propagates. The cosine of this angle tells us what portion of the true velocity is actually measured by the Doppler shift. Normal incidence indicates that the direction of blood flow is at 90° to the direction of propagation of the sound beam. The cosine of 90° is zero. Therefore, zero percent of the true velocity will be measured and the Doppler shift will be absent.

372. Two ultrasound transducers are used to perform Doppler exams on the same patient. The exams are identical except that the transducer frequencies are 5 and 2.5 MHz. Which exam will measure the highest Doppler shift?
   a) the 2.5 MHz exam   b) the 5 MHz exam
   c) neither   d) it cannot be determined

373. Two ultrasound transducers are used to perform Doppler exams on the same patient. The exams are identical except that the transducer frequencies are 5 and 2.5 MHz. Which exam will measure the highest velocities?
   a) the 2.5 MHz exam   b) the 5 MHz exam
   c) neither   d) it cannot be determined

With regard to continuous wave Doppler, are the following four statements true or false?

374. There is a minimum of two distinct piezoelectric crystals in the transducer.

375. Problems with aliasing significantly limit its clinical utility.

376. Doppler shifts measured at the transducer could have been produced from many different locations along the ultrasonic beam.

377. The duty cycle of the wave is 100%.

378. True or False? The appearance of negative velocities in a pulsed Doppler exam always indicates that red blood cells are moving away from the transducer.

379. True or False? The appearance of negative velocities on the spectral display of a continuous wave Doppler exam always indicates that red blood cells are moving away from the transducer producing the sound wave.

372. **B.** When all other things are identical, the use of a higher frequency transducer will produce a greater Doppler shift from blood cells traveling at a particular velocity. Using lower frequency ultrasound will result in lower Doppler shifts. So, for a specific velocity, the Doppler shift produced will depend on the frequency of the incident sound wave, with higher frequencies yielding higher Doppler shifts.

373. **C.** The blood cell velocities measured by Doppler exam are the same, regardless of the frequency of the US signal. Although the frequency shifts associated with the velocities may be different, depending on the frequency of the ultrasound signal, the measured velocities are identical.

374. **True.** With continuous wave ultrasound, one crystal is constantly producing the acoustic wave. A second crystal is required to listen for reflections. In some pulsed ultrasound systems, the same crystal may be responsible for both pulsing and receiving.

375. **False.** Aliasing is an artifact in which high Doppler shifts are misidentified as flow in the opposite direction. Aliasing is produced in pulsed Doppler exams when the PRF is low in comparison to the Doppler frequency. Aliasing does not occur in continuous wave Doppler exams.

376. **True.** One of the characteristics of continuous wave Doppler is that the Doppler shifts received by the transducer can come from anywhere along the length of the acoustic beam. This is called range ambiguity, which is the inability to distinguish the absolute location of the reflectors that produced the Doppler shift.

377. **True.** The duty cycle is the percentage of time that the US transducer is producing a sound wave. With continuous wave Doppler, one crystal is transmitting continuously and the duty cycle is 100% or 1.0.

378. **False.** The appearance of negative velocities on the spectral display of a pulse Doppler exam may indeed indicate that RBCs are traveling away from the transducer. However, aliasing may also cause negative velocities to appear in the spectrum. Aliasing occurs when very high Doppler shifts are incorrectly processed as flow in the opposite direction.

379. **True.** Aliasing does not occur with continuous wave Doppler. Negative velocities appear on the spectral display only when red blood cells travel away from the transducer.

380. In a pulsed Doppler exam, what term is used to describe a very high positive Doppler shift that is displayed as a negative waveform?
    a) attenuation   b) filtering
    c) demodulation   d) aliasing

381. The frequency that defines the highest Doppler shift without the appearance of aliasing is called the _____ and is equal to _____ .
    a) aliasing limit, half of the emitted frequency
    b) pulse repetition frequency, the pulse repetition frequency
    c) Nyquist limit, half of the emitted frequency
    d) Nyquist limit, half of the emitted PRF

382. True or False? In a pulsed Doppler exam, the higher the frequency of the sound emitted by a transducer, the more likely aliasing will occur.

383. True or False? The higher the pulse repetition frequency of a Doppler exam, the more likely aliasing will occur.

384. An ultrasound system has three transducers with frequencies of 7.5 MHz, 5 MHz, and 3.25 MHz. Aliasing appears when the 5.0 MHz transducer is used during a pulsed Doppler exam. What should the sonographer do?
    a) use the 3.25 MHz transducer
    b) use the 7.5 MHz transducer
    c) nothing; the other transducers do not provide solutions to the problem

385. The frequency of a pulsed Doppler wave is 6 MHz, and the pulse repetition frequency is 5 kHz. What is the maximum Doppler shift that can be recorded without aliasing?
    a) 6 MHz   b) 5 kHz
    c) 3 MHz   d) 2.5 kHz

386. During a pulsed Doppler exam, aliasing is noted by the sonographer. Which of the following will not decrease the likelihood of aliasing?
    a) selection of another imaging view with a shallower sample volume
    b) selection of another transducer with a lower frequency
    c) selection of another imaging view that provides a greater pulse repetition period
    d) use of a continuous wave system

380. **D.** In pulsed Doppler exams, the phenomenon wherein high velocities are displayed as negative velocities is called aliasing. The term alias means "having a false identity." Aliasing occurs when the Doppler shift exceeds a particular value called the Nyquist limit.

381. **D.** The Nyquist limit is the highest Doppler shift that can be displayed without aliasing. The Nyquist limit is equal to half of the pulse repetition frequency (PRF).

382. **True.** The Doppler shift produced by red blood cells depends on the frequency of the incident ultrasound wave. The higher the incident frequency, the greater the Doppler shift. The statement is true because a higher emitted frequency produces a higher Doppler shift, and a high Doppler shift is more likely to alias.

383. **False.** Aliasing is less likely to occur with high PRF. Aliasing occurs when the Doppler frequency exceeds half of the PRF. If the PRF is high, it is less likely that the Doppler shift will reach a value that is greater that half of the PRF.

384. **A.** A transducer producing ultrasound at a lower frequency will produce lower Doppler shifts. Lower Doppler shifts are less likely to alias. Therefore, one potential solution for aliasing is to use a transducer that produces lower frequency ultrasound waves.

385. **D.** The frequency at which aliasing occurs is called the Nyquist limit and is equal to one half of the pulse repetition frequency. In this example, the pulse repetition frequency is 5 kHz or 5,000 pulses/sec. The Nyquist limit is half of this value, or 2.5 kHz. Aliasing will occur when Doppler frequencies exceed this value.

386. **C.** The actions that will reduce the likelihood of aliasing are: 1) increasing the PRF by selecting a shallower imaging depth, 2) using a transducer with a lower frequency, and/or 3) using a continuous wave Doppler system that cannot alias. If the sonographer selects an imaging view with a greater pulse repetition period, the pulse repetition frequency will decrease, and the likelihood of aliasing will increase.

387. True or False? Color flow Doppler exams tend to have lower temporal resolution than other forms of two dimensional, real time imaging.

388. True or False? Color flow Doppler imaging incorporates pulsed Doppler principles and provides range resolution.

389. True or False? Color flow Doppler relies on pulsed Doppler principles and is immune to aliasing artifact.

390. True or False? On a color Doppler image, red always represents flow toward the transducer, whereas blue indicates flow away from the transducer.

391. True or False? Color flow Doppler systems display anatomical data in gray scale while simultaneously displaying flow information in color.

392. True or False? Absence of color on a color Doppler image always indicates a region of no blood flow.

393. What is meant by the word "analysis" in the term "spectral analysis" of Doppler signals?
    a) building a sophisticated signal from components
    b) building a simplified signal from components
    c) identifying the building blocks or components of a complex signal
    d) measuring a complex signal and then modifying the information
       it contains

387. **True.** Temporal resolution is the ability to accurately distinguish the location of a structure in time. Color flow technology causes the number of pictures produced each second by an ultrasound system to drop to a very low level. Color flow Doppler suffers from poor temporal resolution because of this reduced frame rate.

388. **True.** One of the greatest advantages of color flow Doppler is its ability to superimpose a two-dimensional color image of velocities upon a two-dimensional gray scale image of anatomy. Therefore, it provides range resolution and is based on pulsed Doppler principles.

389. **False.** The first portion of the statement is true: color flow does indeed rely on pulsed Doppler principles. However, the second part of the statement is false. It is continuous wave Doppler, not pulsed Doppler, that is immune to aliasing.

390. **False.** Although this statement may be true, it is not always true. The color associated with flow toward the transducer appears in the upper half of the color map found on the image. The color associated with flow away from the transducer appears in the lower portion of the color map.

391. **True.** This is the fundamental process by which color flow mapping works. A two dimensional black-and-white image provides data on anatomical structures. A two-dimensional color image is displayed simultaneously on top of the black-and-white image. The color data deals with the Doppler frequencies produced by blood cells flowing in the imaging plane.

392. **False.** Although this statement may be true, it is not always true. Even with blood flow, if the direction of the sound beam is $90°$ to the direction of blood flow, no Doppler shift will be created. The region will remain colorless even though blood flow is present.

393. **C.** The term analysis defines the act of scrutinizing an object to determine its individual components. In Doppler analysis, the complex signal is broken down into its more simple ingredients.

394 The analysis of complex Doppler information using _____ has resulted in the most accurate representation of these spectra.
   a) zero-crossing detectors   b) fast Fourier transforms
   c) time interval histograms   d) Chirp-Z transforms

395. The spectral analysis of color flow Doppler information is most commonly performed by which of the following techniques?
   a) zero-crossing detectors   b) fast Fourier transforms
   c) autocorrelation   d) Chirp-Z transforms

396. The two most common color maps used in color flow imaging are:
   a) variance and direct   b) variance and velocity
   c) turbulent and variance   d) power and velocity

397. True or False? Variance mode color maps may be identified by side-to-side changes in the color bar, whereas velocity mode color maps only change color horizontally.

398. True or False? Velocity mode color Doppler and variance mode color Doppler will produce identical images when blood flow patterns are laminar.

399. True or False? Velocity mode color Doppler and variance mode color Doppler will produce identical images when blood flow patterns are turbulent.

400. Which display is limited to a single pair of brightnesses or display levels?
   a) CRT   b) binary
   c) bistable   d) gray scale

401. Which of the following is not considered a component of an ultrasound system?
   a) master synchronizer   b) pulser
   c) receiver   d) image intensifier

402. True or False? The pulser of a mechanical transducer is typically more complex than the pulser of a phased array transducer.

394. **B.** The most accurate method used to analyze Doppler spectra is called fast Fourier transform, FFT. Techniques such as zero-crossing, time interval histograms and Chirp-Z were initially used to assess Doppler spectra, but were made obsolete by the FFT. Fast Fourier analysis became popular when computers were incorporated into US systems.

395. **C.** Autocorrelation is the technique of choice for spectral analysis of color Doppler information. Although the most accurate technique is the FFT, autocorrelation is better suited to process the enormous amount of data acquired in a color Doppler study.

396. **B.** Two common modes of color Doppler are velocity and variance.

397. **True.** Variance mode maps change colors from side to side, as well as from top to bottom. Velocity mode maps only change color from top to bottom.

398. **True.** The principal difference between velocity and variance modes is that variance mode maps are able to identify turbulent flow. Turbulent flow appears on the image as a third color, such as green, that, on the color map, is located on the right side. If no turbulent flow is present, then the third color will be absent from the image, and the velocity and variance mode images will appear identical.

399. **False.** The principal difference between velocity and variance modes is that variance mode maps are able to identify turbulent flow. Turbulent flow appears on the image as a third color, such as green, that is located on the right side of the color map. When turbulent flow exists, the variance color will appear on the image. Since there is no variance color with velocity mode, the variance and velocity mode images will differ.

400. **C.** A display that presents data in one of only two colors is *bistable*. Each dot on the image is either white or black, bright or dark, on or off.

401. **D.** An ultrasound system is considered to have six components: transducer, pulser, receiver, image processor, display, and master synchronizer. An image intensifier is part of an x-ray imaging system.

402. **False.** A single electrical signal excites the PZT crystal of a mechanical transducer. With phased array technology, a large number of crystals is excited almost simultaneously to produce an acoustic pulse. This technique is used by phased array systems to steer and focus ultrasound beams. Therefore, phased array systems have more complex pulsers.

403. What typical pulser output voltage excites a piezoelectric crystal?
   a) 0.1 volts   b) 500 mvolts
   c) 110 volts   d) 10 µvolts

404. Which of the following tasks is not performed by the receiver of an
   ultrasound system? (More than one answer may be correct.)
   a) rectification   b) smoothing
   c) compression   d) degaussing

405. What is the typical voltage of the input signal to the receiver of an
   ultrasound system?
   a) 0.1 volts   b) 500 mvolts
   c) 250 volts   d) 10 µvolts

406. The output gain of an ultrasound pulser determines the _____ of the
   acoustic pulse.
   a) imaging depth   b) intensity
   c) duration   d) pulse repetition period

407. Which of the following terms does not belong this group?
   a) receiver gain   b) energy output
   c) pulser power   d) acoustic power
   e) transmitter output   e) output gain

408. True or False? The sonographer can alter the pulser power.

409. Which of the following functions are performed by the receiver of an
   ultrasound system? (More than one answer may be correct.)
   a) amplification   b) threshold
   c) compensation   d) demodulation

410. Which of the following is a typical value for the amplification of a signal
   by the receiver?
   a) 50 to 100 watts   b) -50 to -100 dB
   c) 50 to 100 dB   d) 5 to 25 W/cm^2

411. True or False? The sonographer can adjust the receiver gain.

403. **C.** The signal typically produced by the pulser of an ultrasound system, used to excite the piezoelectric crystal of a transducer, may be in the range of hundreds of volts.

404. **D.** Degaussing is the only task listed above that is not performed by the system's receiver.

405. **D.** The signal produced by the transducer upon reception and sent to the receiver of the ultrasound system is extremely small and is in the micro- to millivolt range.

406. **B.** The output power of the pulser determines the strength or the intensity of the acoustic pulse produced by the transducer. The greater the electrical voltage from the pulser that strikes the piezoelectric crystal of the transducer, the greater the intensity of the acoustic pulse.

407. **A.** All of the terms listed, except for choice A, refer to the strength of the signal produced by the transducer. Receiver gain is not synonymous with the other five choices.

408. **True.** The output voltage of the pulser and the strength of the acoustic pulse produced by the transducer are in the direct control of the sonographer. The output power is assigned different names by the various ultrasound manufacturers, but the sonographer can always alter its level.

409. **A, B, C and D.** All of the functions listed above are performed by the ultrasound system's receiver.

410. **C.** The amplification of the weak electrical signals that first reach the receiver ranges from 50 to 100 decibels. This prepares the signals for further processing by the receiver and other system components.

411. **True.** The receiver gain is in direct control of the sonographer. The degree to which the signal returning from the transducer is amplified is determined and can be adjusted by the sonographer. It is a primary control on the ultrasound system that determines the ultimate quality of images produced during an examination.

412. What is the term for adjusting for path length-related attenuation?
    a) compression   b) compensation
    c) GTC   d) reconfirmation

413. Which does not belong in the following terms that describe compensation?
    a) swept gain   b) depth gain
    c) time gain   d) amplitude gain

414. The region of minimum amplification on a standard TGC curve is related
    to the _____.
    a) far field   b) focus only
    c) area close to the transducer   d) focal zone

415. Which transducer system is most likely to have the longest delay in its
    TGC curve?
    a) a 5 MHz linear array   b) a 2.5 MHz annular array
    c) a 7.5 MHz mechanical   d) a 5 MHz annular array

416. In which region of the TGC curve is the slope of the swept gain
    compensation most likely to contribute to improved image quality in a
    diagnostic scan?
    a) the region very close to the transducer
    b) the far zone
    c) the focal zone

417. True or False? The lower the frequency of the transducer, the steeper the
    slope of the TGC curve.

418. True or False? The lower the frequency of the ultrasound beam, the
    shallower the delay of the TGC curve.

419. What does the far gain of a depth gain compensation curve represent?
    a) the median amplification related to compensation
    b) the maximum amplification related to compensation
    c) the minimum amplification related to compensation
    d) the maximum attenuation related to compensation

412. **B.** Ultrasound attenuates as it travels in soft tissue. Therefore, the deeper sound travels, the lower the reflected pulse's intensity. The process of adjusting for this path length attenuation is called compensation.

413. **D.** The terms swept gain, depth gain, and time gain are used interchangeably when describing compensation. The term amplitude gain is not synonymous with the other three terms.

414. **C.** The region of minimum amplification on a TGC curve, often called the "delay," is always associated with the area close to the transducer.

415. **B.** The 2.5 MHz transducer will be more likely to have a long delay in the TGC curve. Low frequency ultrasound waves attenuate to a lesser extent than high frequency waves. Since choice B has the lowest frequency, it will have the least amount of attenuation and therefore will require less compensation in the near zone. A long delay is consistent with little compensation in the region close to the transducer. The type of transducer, (e.g., annular, linear) does not typically affect the TGC.

416. **C.** TGC is most effective in improving image quality in the focal zone. Compensating for attenuation in the focal zone will allow the ultrasound system to produce high-quality scans with detailed information.

417. **False.** With lower frequency transducers, there is less need for a steep slope in the TGC curve. In comparison to high-frequency sound, the TGC curve can be less steep with low-frequency ultrasound because low frequencies do not attenuate as rapidly in soft tissue. Therefore, they do not require the extra amplification provided by the compensation process.

418. **False.** With lower frequencies of sound, the delay of the TGC curve is deeper. There is less attenuation with low frequency sound, and this allows for a deeper TGC delay.

419. **B.** The far gain setting of a TGC curve represents the maximum amplification that a reflected wave undergoes during the process.

420. The relationship between the largest and the smallest signal amplitudes
processed by an instrument is called the _____ and is expressed
in units of _____.
   a) decibels, watts   b) intensity, W/cm^2
   c) dynamic range, dB   d) proportionality, dB

421. Why is it necessary for the receiver to perform compression on the
electrical signals it processes?
   a) the dynamic range of the system's electronics is less than the dynamic
      range of the received ultrasound reflection
   b) the dynamic range of the display is greater than the dynamic range of
      the transducer
   c) the dynamic range of the receiver is less than the dynamic range of the
      image processor
   d) the dynamic range of the display exceeds the dynamic range of all
      other components of the ultrasound system

422. True or False? Typically, the sonographer can adjust the compression rate
of a receiver.

423. Which of the following terms cannot be used to describe the process of
eliminating low amplitude signals from further processing?
   a) reject   b) subordination
   c) threshold   d) suppression

424. True or False? Typically, the sonographer can adjust the threshold level
of a receiver.

425. Which of these components typically has the greatest dynamic range?
   a) display   b) pulser
   c) amplifier   d) demodulator

426. Which of the following components typically has the lowest dynamic
range?
   a) display   b) pulser
   c) amplifier   d) demodulator

420. **C.** The dynamic range is defined as the ratio of the largest to the smallest signal amplitude that a device can process. The units of dynamic range are decibels: dB.

421. **A.** The process of compression decreases the dynamic range of a signal. Ultrasound systems compress signals because the dynamic range of the reflected signals received by the transducers is far greater than the dynamic range of any of the other components of the system. Thus, compression matches the dynamic range of the signal to the dynamic range of the system's electrical components.

422. **False.** The degree of compression of a signal processed in an ultrasound system is usually determined by the system electronics and cannot be changed or adjusted by the operator.

423. **B.** The process of removing low-level signals from further processing or display by an ultrasound system is called reject, threshold, or suppression. Subordination does not describe this procedure.

424. **True.** Under normal circumstances, the operator adjusts the reject (suppression or threshold) level that is performed during signal processing.

425. **C.** Of the four choices, the component of an ultrasound system with the greatest dynamic range is the amplifier.

426. **A.** Of the four choices, the component of an ultrasound system with the lowest dynamic range is the display.

427. Which of the following tasks is incorporated in the process of
demodulation? (More than one answer may be correct.)
    a) smoothing   b) amplification
    c) rectification   d) decompression

428. True or False? The primary purpose of demodulation is the preparation of
the electrical signal for display on a television.

429. True or False? Typically, the sonographer can control the demodulation
process performed by a receiver.

430. Which of the following is an analog number?
    a) the weight of a person   b) the number of people in a room
    c) the number of stars in the sky   d) the number of tires on a car

431. True or False? A digital representation of a number can only achieve
specific fixed values.

432. What was the original role of an ultrasound system's analog scan
converter?
    a) to increase the dynamic range of ultrasound systems
    b) to make real-time imaging possible
    c) to increase the sensitivity of ultrasound systems
    d) to allow for gray scale imaging

433. Which of the following terms isn't a limitation of analog scale converters?
    a) image fade   b) low spatial resolution
    c) image flicker   d) tube deterioration

434. True or False? Both digital and analog scan converters must use computer
technology to process electronic data for image display.

435. What is the smallest element in a digital picture called?
    a) the bit   b) the byte
    c) the pixel   d) the fractel

436. What is the name for the smallest amount of digital storage?
    a) the bit   b) the byte
    c) the pixel   d) the fractel

427. **A and C.** Demodulation is a two-step process. These steps are rectification and smoothing.

428. **True.** The primary purpose for the demodulation of a signal in an US system is to prepare it for ultimate display on a television monitor.

429. **False.** With the majority of ultrasound systems, the process of demodulation is not under the control of the sonographer. The parameters of demodulation are set within the ultrasound system and cannot be adjusted by the sonographer.

430. **A.** The weight of a person can be considered an analog number because it is continuous and can achieve an unlimited number of values. On the other hand, the number of people in a room or the number of tires on an automobile can only be represented by discrete, whole, or integer values.

431. **True.** A digital representation of a number is limited to discrete values and cannot take on unlimited values. Because of this, digital scan converters can display only a fixed number of gray shades.

432. **D.** The introduction of scan converter technology to ultrasound allowed for the presentation of gray-scale images. That is, the images were not limited to black and white (bistable), but had various shades of gray within them.

433. **B.** Analog scan converters have very fine spatial resolution. It is common for an analog converter to divide a picture into more than 1 million very small picture elements. However, the other three selections are indeed problems or limitations associated with their use.

434. **False.** Digital scan converters must use computer technology to process data. This is not necessary in analog scan converters.

435. **C.** The smallest element in a digital representation of an image is the *pixel*. The term is derived from the words "picture element."

436. **A.** The *bit* is the smallest amount of digital storage or computer memory. The term bit is derived from the words "binary digit." A combination of eight bits makes a *byte*. Computer memory is also called random access memory or RAM.

437. True or False? The greater the number of bits assigned to each pixel of a digital image, the greater the spatial resolution of the image.

438. True or False? The greater the number of bits assigned to each pixel of digital image, the greater the number of shades of gray in the image.

439. True or False? In order for a digital image to be displayed on a television screen, the data must be processed by a digital to analog converter.

440. True or False? Preprocessing of image data occurs after the data has been stored in the scan converter.

441. True or False? Postprocessing of image data occurs after the data has been stored in the scan converter.

442. Two digital scan converters are undergoing evaluation. Both produce images of the same size. System A has 1,000,000 pixels, with 10 bits assigned to each. System B has 250,000 pixels, with 12 bits assigned to each. Which system is more likely to have the capability of displaying very small details in an image?
a) System A   b) System B
c) both are the same   d) it cannot be determined

443. Two digital scan converters are undergoing evaluation. System A has 1,000,000 pixels, with 10 bits assigned to each. The picture from this system is 100 square inches. System B, with an image size of 10 square inches, has 250,000 pixels, with 12 bits assigned to each. Which system is more likely to have the capability of displaying very small details in an image?
a) System A   b) System B
c) both are the same   d) it cannot be determined

444. Two digital scan converters are undergoing evaluation. System A has 1,000,000 pixels, with 4 bits assigned to each. System B has 250,000 pixels, with 12 bits assigned to each. Which system is more likely to have the capability of displaying very subtle differences in gray scale?
a) System A   b) System B
c) both are the same   d) it cannot be determined

437. **False.** The number of bits assigned to each pixel of a digital image does not directly affect the spatial resolution of the image.

438. **True.** The most dramatic and obvious change in a digital picture when the number of bits assigned to each pixel in the image increases is that the number of shades of gray increases.

439. **True.** The representation of an image in digital format is not suitable for display on a standard television because a standard television is an analog device. Therefore, after all of the signal processing is completed, the image data is passed through a digital to analog converter. It is then suitable for display on the television.

440. **False.** Preprocessing of image data is achieved prior to image storage in the scan converter. Amplification and write zoom are examples of preprocessing.

441. **True.** All processing that occurs after the image is stored in the scan converter is called postprocessing. Examples of postprocessing are read magnification and the various forms of gray scale assignment. Note that any manipulation of a frozen image is postprocessing.

442. **A.** The scan converter with the greatest number of pixels assigned to an image of fixed size will provide the viewer with the greatest spatial detail. In this case, System A has a higher pixel density and, therefore, the ability to display greater detail.

443. **B.** System B has a greater ability to display small detail in an image in comparison to System A. The reason is that System B has 25,000 pixels per square inch, while System A only has 10,000 pixels per square inch. System B has a greater pixel density, and this determines the degree of spatial resolution.

444. **B.** Gray scale levels are determined by the number of bits assigned to each pixel in an image. Since System B has 12 bits assigned to each pixel, while System A has only 4, System B will display more gray shades and, thus, display subtle differences.

445. Two ultrasound systems are undergoing evaluation. System A has 1,000,000 pixels, with 4 bits assigned to each. System B has 750,000 pixels, with 12 bits assigned to each. Which system has the best longitudinal resolution?
    a) System A   b) System B
    c) both are the same   d) it cannot be determined

446. A video display that is limited to only black and white, with no other shades of gray, is called _____ .
    a) binary   b) bistable
    c) monochrome   d) unichrome

447. Which of the following terms is not included in information processed by a standard duplex ultrasound system?
    a) attenuation   b) amplitude
    c) frequency   d) time of arrival

448. In a standard cathode ray tube used to display ultrasound images, what are the charged particles that are emitted from a "gun" from the rear of tube?
    a) electrons   b) positrons
    c) photons   d) neutrons

449. What is the coating on the inner surface of a cathode ray tube that glows when it is struck by charged particles?
    a) sulfur   b) phosphor
    c) tungsten   d) carbon

450. How does a cathode ray tube steer the charged particles emitted by the electron gun, located in the back of the tube, so that they sweep across the front screen?
    a) by phasing the source
    b) by mechanically steering the gun
    c) by sending the charged particles through a time-varying magnetic field
    d) by reflecting the particles from a mirrored surface

445. **D.** The longitudinal resolution of an ultrasound system is determined primarily by the pulse duration. Since no information is provided for pulse duration, it is impossible to determine which of the systems will have superior longitudinal resolution.

446. **B.** A video display with the ability to show only black and white, without any intermediate shades of gray, is called bistable.

447. **A.** A duplex ultrasound system provides both image and Doppler information. Included in the data processing is the amplitude of the reflected signal, its time of arrival, and its frequency. The pulse's attenuation is not recorded or processed.

448. **A.** The end opposite to the screen of a CRT has a small emitter of electrons. The coating on the inner surface of the screen glows when struck by these electrons and provides the capacity to produce an image.

449. **B.** The inner surface of a CRT is coated with a phosphor that glows when it is struck by electrons. This allows images to be produced on the screen. Color televisions have three different phosphor coatings on the screen. Each phosphor glows with a different color when excited by electrons, thereby producing a color image.

450. **C.** The electrons that are shot from the gun mounted in the back of a cathode ray tube are negatively charged particles. Like little metallic objects, these electrons will react when they are placed in a strong magnetic field. The electrons are swept across the face of the CRT by forcing them to travel through a strong magnetic field that varies rapidly in time.

451. How many horizontal scan lines make up an image on a television monitor used in the United States?
   a) 2,500
   b) 500
   c) 525
   d) 1,000

452. A standard television system processes information rapidly to provide the viewer with the sensation of a real-time display. How is this achieved?
   a) 60 frames/sec in sequence
   b) alternate frames at a rate of 30/sec
   c) alternate even and odd fields at a rate of 60/sec
   d) even and odd fields at a rate of 60 frames/sec

453. What is the approximate number of frames that must be presented each second for the human eye to perceive the display without flickering?
   a) 2
   b) 15
   c) 60
   d) 500

454. What is the maximum number of shades of gray that can be portrayed with 4 bits?
   a) 4
   b) 8
   c) 12
   d) 16

455. With 6 bits, what is the largest number of different gray shades that can be stored?
   a) 8
   b) 16
   c) 64
   d) 256

456. What occurs when the number of bits apportioned to each pixel of an image is increased?
   a) the greater the spatial detail   b) there is greater temporal resolution
   c) the larger the field of view   d) there are more shades of gray

457. What happens to a digital image when the pixel density increases?
   a) there is more spatial detail   b) the temporal resolution increases
   c) the larger the field of view   d) there are more shades of gray

451. **C.** In the United States, a standard television picture comprises 525 horizontal lines. The electron beam is rapidly steered across the face of the CRT from left to right and from top to bottom and "paints" an image.

452. **C.** Standard television within the United States uses an interlaced display. With an interlaced display, the 213 odd-number lines are painted onto the screen in order. This is called the odd field. Subsequently, the 212 even-numbered lines, composing the even field, are drawn on the screen. A single frame is made of 1 even and 1 odd field. A total of 30 frames are displayed in a second. Each frame has an odd and an even field; hence, 60 fields are presented each second.

453. **B.** Approximately 15 frames/sec are required to perceive motion in a smooth and flicker-free fashion. When the frame rate is significantly below this value, the viewer experiences flicker. Flicker increases fatigue and discomfort and diminishes the observer's perceptive ability.

454. **D.** A maximum of 16 shades of gray can be represented with 4 bits. To calculate this, multiply 2 by itself in times equal to the number of bits. In this example, multiply 2 by itself a total of 4 times (there are 4 bits): 2 x 2 x 2 x 2 = 16 shades.

455. **C.** A maximum number of 64 shades can be stored with the use of 6 bits. We multiply 2 by itself a total of 6 times to calculate the answer to this question: 2 x 2 x 2 x 2 x 2 x 2 = 64 shades.

456. **D.** In a digital image, as the number of bits assigned to each pixel increases, the number of different shades of gray in the image increases.

457. **A.** The pixel density, the number of pixels per inch, determines the detail that a digital picture can illustrate. The greater the pixel density, the higher the spatial detail. The fewer pixels per inch, the lesser the detail.

458. How many binary digits are required to store 29 different levels of gray?
    a) 4   b) 5
    c) 6   d) 29

459. All of the following are true of read zoom except _____.
    a) it is a preprocessing function of the receiver
    b) the same number of pixels appear in the original region of interest
        (ROI) and the and the zoomed image
    c) on the screen, the pixels in the zoomed image are larger than those in
        the region of interest (ROI)
    d) the same number of ultrasound pulses are used to create the original
        ROI and the zoomed image

460. All of the following are true of write zoom except _____.
    a) it is a preprocessing function of the receiver
    b) the same number of pixels appear in the original ROI and the zoomed
        image
    c) on the screen, the pixels in the zoomed image are the same size as
        those in the ROI
    d) in comparison to the zoomed image, fewer ultrasound pulses are used
        to create the original ROI

461. True or False? The assignment of different shades of gray to the digital
numbers stored in the scan converter acts to improve the diagnostic value
of an ultrasound exam.

462. Acoustic artifacts include _____. (More than one answer may be correct.)
    a) images of reflectors in an inappropriate position
    b) images that have reflectors of improper shape
    c) images of reflectors of incorrect brightness
    d) images that do not correspond to anatomical structures

463. Ultrasound systems are designed to automatically _____.
    a) display similar structures with identical intensities, regardless of depth
    b) position structures at the correct depth, regardless of the medium
    c) display all reflections on a line corresponding to the major axis of the
        ultrasound beam, regardless of refraction
    d) select the optimal beam width based on the clinical application

458. **B.** A minimum of 5 bits is required to display 29 shades of gray. Actually, 5 bits can display a maximum of 32 shades of gray. Because only 16 shades of gray can be represented by 4 bits, a fifth bit must be added to achieve adequate memory to display 29 shades of gray.

459. **A.** Read zoom is a postprocessing function of the receiver. The data originally in the scan converter, before zoom, remains intact. New information is not acquired.

460. **B.** Write zoom is a preprocessing function of the receiver where new information is acquired from the region of interest. The data originally in the scan converter, before zoom, are discarded. All of the pixels in the image are now concentrated into the ROI, thereby increasing the pixel density and image quality of the zoomed portion.

461. **True.** By using either high-level or low-level gray scale enhancement, the sonographer selects the tissue textures that will be distinguished from each other on the image. This is very important when different tissues in the body produce similar reflections. By using an appropriate gray scale enhancement scale, these tissues will appear differently on the screen, thereby enhancing the diagnostic power of the exam.

462. **A, B, C and D.** An artifact is an error in imaging. All of the choices indicate a type of error and therefore are artifacts.

463. **C.** Ultrasound systems assume that all reflectors lie directly along the main axis of the ultrasound beam. An artifact is created when pulses refract and change direction.

464. Which of the following is not a potential cause of artifact in diagnostic imaging?
   a) operator error   b) equipment malfunction
   c) patient motion   d) ultrasound physics   e) none of the above

465. What is true of artifacts related to depth resolution?
   a) too many reflectors are displayed on the image
   b) position reflectors are too deep on the image
   c) too few reflectors are on the image
   d) position reflectors are displayed too shallow on the image

466. Axial resolution artifacts are due to which of the following?
   a) multiple reflections   b) beam width
   c) attenuation   d) pulse length

467. Which of the following pulses would be least likely to produce a radial resolution artifact?
   a) 10 MHz, 8 mm beam diameter, 4 cycles per pulse
   b) 4 MHz, 4 mm beam diameter, 2 cycles per pulse
   c) 7.5 MHz, 8 mm beam diameter. 2 cycles per pulse
   d) 6 MHz, 2 mm beam diameter, 2 cycles per pulse

468. Lateral resolution artifacts would be most profound with which of the following systems?
   a) 10 MHz, 4 mm beam diameter, 4 cycles per pulse
   b) 4 MHz, 4 mm beam diameter, 2 cycles per pulse
   c) 7.5 MHz, 8 mm beam diameter. 2 cycles per pulse
   d) 6 MHz, 2 mm beam diameter, 2 cycles per pulse

469. The lateral resolution of four ultrasound machines is listed below. Which of the systems would produce the finest quality picture?
   a) 2 cm   b) 4 mm
   c) 6 mm   d) 8 hm

470. Two small cysts, positioned perpendicular to the main axis of an ultrasound beam, are 2.4 mm apart. What determines whether these structures will appear as two distinct images on the system's display?
   a) the beam width   b) the pulse length
   c) the PRF   d) the TGC

464. **E.** The four choices, A through D, represent the primary explanations of artifacts commonly seen in diagnostic ultrasound.

465. **C.** Depth resolution is the ability to distinguish two reflectors that lie close to each other along the main axis of the beam. Most errors in depth resolution result in fewer reflections placed on the image than are actually in the body.

466. **D.** The critical factor that influences axial resolution is pulse length. Shorter pulses produce higher quality images than do longer pulses.

467. **C.** Short pulses are least likely to produce a radial resolution artifact. Short pulses are those with a high frequency and few cycles per pulse. Although choice C does not have the lowest frequency, it only has 2 cycles per pulse and therefore is the shortest of the 4 pulses.

468. **C.** Lateral resolution artifact is related to the diameter of the ultrasound beam. Therefore, choice C, with a beam diameter of 8 mm, is most likely to produce a lateral resolution artifact.

469. **B.** Resolution is reported in units of distance, with smaller values producing higher quality pictures. Choice B is the smallest and therefore represents the system with the best images based on lateral resolution.

470. **A.** The ability to correctly image structures that lie close to each other in a direction perpendicular to the main axis of the ultrasound beam is called lateral resolution. Lateral resolution is primarily determined by the width of the acoustic pulse.

471. In standard diagnostic imaging instrumentation, which has the higher numerical value?
    a) longitudinal resolution   b) lateral resolution
    c) neither, they have identical values

472. What produces the artifact known as acoustic speckle?
    a) refraction   b) attenuation
    c) interference of tiny acoustic wavelets
    d) resonance of particles in the near field

473. What is true of acoustic speckle?
    a) its effects are mild and it tends to slightly degrade images
    b) it is a rare artifact that does not occur in real-time imaging
    c) its effects are profound
    d) nothing can be done to correct it

474. What artifact results from an ultrasound beam having a finite and measurable three-dimensional profile?
    a) acoustic speckle   b) multipath artifact
    c) slice thickness artifact   d) grating lobe artifact

475. How does thickness artifact commonly express itself?
    a) improper location of reflections   b) improper brightness of reflectors
    c) absence of reflectors   d) strong linear echoes

476. Which artifact results in improper side-by-side positioning of reflectors?
    a) multipath   b) comet tail
    c) refraction   d) reverberation

477. What assumption is violated when the receiver processes a refracted acoustic wave?
    a) waves travel directly to and from a reflector
    b) sound travels at an average speed of 1.54 mm/us
    c) sound travels in a straight line
    d) the acoustic imaging plane is very thin

478. Six distinct, equally spaced reflections appear on an image at ever increasing depths. What type of artifact does this scenario describe?
    a) reverberation   b) ring down
    c) mirror imaging   d) longitudinal resolution

471. **B.** In terms of image quality, longitudinal resolution is usually better than lateral resolution. Thus, the number assigned to lateral resolution is higher than that assigned to longitudinal resolution.

472. **C.** Acoustic speckle is produced by the constructive and destructive interference of small acoustic waves reflected from small particles in the near field.

473. **A.** Even though it does not accurately represent reflectors in the medium, acoustic speckle is often considered "tissue texture." It tends to contribute to overall image degradation.

474. **C.** Slice artifact occurs because an ultrasound beam has a thickness. Although the sonographer may perceive the imaging plane as being extremely thin, the beam actually has a measurable thickness. Thickness artifact results in some reflectors appearing in the image even though they are positioned on either side of the idealized imaging plane.

475. **A.** Slice thickness artifact results in reflections on the image that do not correlate with the anatomical position of the reflector.

476. **C.** Refraction artifact results from the acoustic pulse bending and changing direction. The ultrasound system cannot correct this change, which results in an improper lateral location of reflectors on the image.

477. **C.** Ultrasound systems are designed on the basis of certain assumptions. Artifacts occur when these assumptions are invalid. Refraction artifact results from sound traveling in a crooked or bent line while the ultrasound system assumes a straight path.

478. **A.** Reverberations are multiple, equally spaced reflections.

479. Which situation will commonly produce reverberation artifact?
    a) two masses that lie perpendicular to the main axis of the sound beam
    b) two weak reflectors that lie close to each other along the axis of
        the beam
    c) two strong reflectors that lie along the main axis of the beam
    d) a single mass of highly reflective material

480. Mirror imaging artifact is created by which of the following?
    a) reflection   b) refraction
    c) propagation speed error   d) attenuation

481. The mirror image artifact is positioned _____.
    a) shallower on the image than in the body
    b) always deeper on the image than in the body
    c) sometimes the same depth on the image as in the body
    d) never deeper on the image than in the body

482. What are the characteristics of a medium that produces comet tail artifact?
    a) weak reflectors, closely spaced, low propagation speeds
    b) strong reflectors, widely spaced, high propagation speeds
    c) strong reflectors, closely spaced, low propagation speed
    d) strong reflectors, closely spaced, high propagation speed

483. What is a comet tail artifact's fundamental mechanism of formation?
    a) reflection   b) rarefaction
    c) refraction   d) redirection

484. In general diagnostic imaging, what is the primary effect of multipath
    artifact on an image?
    a) poor angular resolution   b) acoustic speckle
    c) mild image degradation   d) gross horizontal misregistration

485. Side lobe artifact results from which of the following?
    a) sound beams bending
    b) linear array transducer architecture
    c) unexpectedly low acoustic attenuation
    d) acoustic energy radiating in a direction other than the beam's main
        axis

479. **C.** Reverberations are commonly produced by a pair of strong reflectors that lie along the main axis of an ultrasound beam.

480. **A.** Mirror imaging is created by an unexpected and uncorrected reflection of the ultrasound pulse from a strong reflector.

481. **B.** A mirror image artifact occurs when a pulse bounces off a strong reflector (such as the diaphragm) and then strikes a second reflector. The pulse's path is lengthened by the mirroring, and the second reflector (artifact) is always displayed at an abnormal great depth.

482. **D.** A comet tail artifact is similar to a reverberation except that the reflections on the image are not distinct. The reflections of a comet tail are closely spaced because of strong reflectors with high propagation speeds lying close to each other.

483. **A.** Comet tail artifacts appear on an image as a result of numerous reflections.

484. **C.** Multipath artifact does not manifest itself in a pronounced manner. The primary result of this artifact is mild image degradation.

485. **D.** A side lobe is produced when a significant amount of acoustic energy is directed along a line different from the main axis of the acoustic pulse.

486. Side lobe artifact usually results in all of the following except _____.
    a) hollow structures appearing "filled in" on the image
    b) reflectors not appearing on an image
    c) reflectors appearing in improper locations on the image
    d) reflectors appearing in multiple locations on the image

487. Grating lobes are most common with which type of transducer technology?
    a) annular array   b) continuous wave
    c) mechanical scanners   d) linear arrays

488. Grating lobes are produced by the same mechanism as which other artifact?
    a) side lobes   b) reverberation
    c) transaxial lobes   d) acoustic speckle

489. True or False? Grating lobes are a result of substantial acoustic energy directed outward from a linear array transducer, but not along the main axis of the sound beam.

490. Which technique of linear array transducer design has virtually eliminated the appearance of grating lobe artifact on modern ultrasound systems?
    a) demodulation   b) subdicing
    c) deconvolution   d) none of the above

491. True or False? Grating lobes are attributed only to linear array transducers.

492. Where are shadowing artifacts commonly seen on an acoustic scan?
    a) distal to a structure with a high impedance
    b) proximal to a structure with a low propagation speed
    c) distal to a structure with a high attenuation
    d) next to a structure with a low elastance

493. How is shadowing artifact expressed?
    a) placing structures too deep on the image
    b) placing structures in improper lateral position
    c) placing reflections in multiple locations
    d) reflectors being absent on the image

486. **B.** It is unlikely that a side lobe artifact would result in the absence of a reflector on an image. It is more likely that additional reflections will appear on the image .

487. **D.** Grating lobe artifacts are associated with linear array transducers.

488. **A.** Side lobes and grating lobes are similar in that they represent acoustic energy transmission in a direction other than the main axis of an acoustic beam.

489. **True.** This is the exact definition of a grating lobe.

490. **B.** The process of subdicing has almost eliminated the appearance of grating lobe artifact on linear array transducer systems. Subdicing consists of dividing each of the crystals in the array into smaller pieces. The subdiced pieces are then fired simultaneously as if they were a single crystal. This process minimizes grating lobes.

491. **True.** Grating lobes are associated with linear array transducers.

492. **C.** Shadowing occurs on an image at locations beneath strong attenuators. For example, a region behind a gallstone may be affected by shadowing.

493. **D.** A shadow artifact results from unexpectedly high attenuation. The ultrasound beam deeper than the strong attenuator is so weak that significant anatomical information may not be presented on the image.

494. Which of the following can produce shadowing?
    a) refraction
    b) multipath
    c) reflection
    d) all of the above

495. Unexpectedly low attenuation results in which of the artifacts listed below?
    a) refraction
    b) attenuation
    c) enhancement
    d) shadowing

496. When enhancement occurs, where does it appear on the image?
    a) in the near field
    b) distal to a weak attenuator
    c) in the far field
    d) proximal to a weak reflector

497. Which artifact may be produced by acoustic focusing of an ultrasound beam?
    a) side lobes
    b) refraction
    c) speckle
    d) enhancement

498. Which of the following artifacts occurs as sound travels through a medium with a propagation speed different from that of soft tissue?
    a) vertical misregistration
    b) horizontal misregistration
    c) lateral resolution
    d) ring down

499. While imaging a test object, an ultrasound machine displays one image 1.8 cm deeper than another. Upon measuring the test object, it is found that one reflector is actually 2.0 cm deeper that the other. What conclusion can be drawn from this?
    a) The propagation speed of the test object is the same as that of soft tissue
    b) The propagation speed of the test object is less than that of soft tissue
    c) The attenuation of the test object is less than that of soft tissue
    d) The propagation speed of soft tissue is less than that of the test object

500. If an ultrasound pulse travels through a large mass in the body at a speed of 1.2 mm/μs, what happens to the position of all echoes produced from reflectors shallower than the mass?
    a) they are placed in too shallow a location on the image
    b) they are placed in too deep a location on the image
    c) they are likely to be placed at the correct depth

494. **D.** Shadowing results from an unexpected weakening of ultrasound pulses. This weakening can occur with refraction, attenuation, and reflection.

495. **C.** Enhancement occurs with unexpectedly low rates of attenuation.

496. **B.** Enhancement occurs after an ultrasound pulse has traveled through a low attenuator and appears on the image deeper than (distal to) that structure.

497. **D.** Focusing of an ultrasound beam may result in very strong reflections from structures in the body. The powerful reflections can be considered enhancement.

498. **A.** Vertical misregistration occurs when pulses do not travel at the speed of sound in soft tissue (1,540 m/sec.) The reflections on the image will be placed at greater depths than the actual position of the reflector in the body.

499. **D.** When vertical misregistration occurs, it is possible that the speed of sound in the medium differs from 1,540 m/sec. If the measurements on the image are less than the true distances, then the speed of sound in the medium is greater than that through soft tissue.

500. **C.** The mass has a propagation speed different from that of soft tissue. Structures deeper than the mass will be affected by this. However, this question asks about reflectors that are shallower than the mass. Images from structures shallower than the mass are not affected by its presence.

501. If an ultrasound pulse travels through a large mass at a speed of 1.2 mm/us, what happens to the position of all echoes produced from reflectors distal to the mass?
    a) they are placed too shallow on the image
    b) they are placed too deep on the image
    c) they are likely to be placed at the correct depth

502. What is the most likely cause of ring down artifact?
    a) refraction   b) inversion
    c) reabsorption   d) resonance

503. What type of artifact causes an ultrasound reflection to be placed at an incorrect depth?
    a) lateral incertitude   b) shadowing
    c) range ambiguity   d) indeterminate relaxation

504. Which artifact is not affected by the shape or structure of an ultrasound pulse?
    a) lateral resolution   b) slice thickness
    c) mirror imaging   d) longitudinal resolution

505. Which artifact is not related to the unexpected reflection of an acoustic wave?
    a) multipath   b) comet tail
    c) reverberation   d) lateral resolution

506. True or False? All artifacts are errors in imaging that do not represent the true anatomy of the imaged organ. Therefore, they are all undesirable and should be eliminated if possible.

507. True or False? Quality assurance evaluations of ultrasound systems should be performed weekly.

508. True or False? Quality assurance evaluations of ultrasound systems can only be performed by biomedical engineers or physical scientists.

509. True or False? Proper quality assurance programs require the assessment of every transducer in the clinical laboratory.

501. **B.** The mass has a slow propagation speed when compared with that of soft tissue. All structures distal to the mass will be placed too deep on the image. The reflections on the image will be farther than the reflector's true anatomical position.

502. **D.** Ring down artifact is thought to result from resonation of a small structure in the medium. This sustained vibration produces a long, linear artifact distal to the resonating structure.

503. **C.** There can be uncertainty regarding the location of a structure that produces a reflection when the pulse repetition frequency of a system is increased. This artifact, which is typical of Doppler systems, is called range ambiguity.

504. **C.** Of the artifacts listed, only mirror imaging is unrelated to the size, shape, or geometry of the ultrasound pulse. The others depend in some way on the pulse's characteristics.

505. **D.** Lateral resolution is an artifact that is not associated with abnormal reflection. Lateral resolution artifacts are related to beam diameter.

506. **False.** All artifacts are indeed errors in imaging. However, they may provide useful information. An artifact can provide insight into the physical characteristics of a structure, and help the clinician make a definitive diagnosis. Eliminating all artifacts may exclude important information from appearing on the exam.

507. **False.** Quality assurance evaluations must be performed *routinely*. The actual time between evaluations is determined by the pattern of use of the system.

508. **False.** Quality assurance evaluations should be performed by the sonographers using the system. Although other individuals may be capable of evaluating system performance, the sonographer plays an essential role in the quality assurance of an ultrasound system.

509. **True.** To establish a valid and rigorous quality assurance program, each component in clinical use must be evaluated. Thus, each transducer must be individually assessed.

510. Which of the following statements are true of a tissue equivalent phantom? (More than one answer may be correct.)
    a) it has the same propagation speed as that of soft tissue
    b) it attenuates ultrasound at a rate similar to that of soft tissue
    c) embedded within it are solid masses and cystic structures
    d) it is the device of choice for quality assurance studies

511. Which two of the following devices convert ultrasound energy into heat?
    a) a calorimeter   b) a thermocouple
    c) a hydrophone   d) a Schlieren

512. True or False? The biological effects of ultrasound are thought to be negligible and, therefore, few investigations have been performed on the subject.

513. True or False? It is generally believed that the effects of ultrasound on biologic media are minimal.

514. True or False? It is generally believed that the biological effects of ultrasound are minimal at intensity levels typical of those produced by diagnostic imaging equipment.

515. Which of the following is not considered a potential mechanism for the production of bioeffects from ultrasound exposure to the body?
    a) temperature elevation   b) fractionation
    c) cavitation   d) mechanical trauma

516. A comprehensive and scholarly review of bioeffects was performed by the American Institute of Ultrasound in Medicine for all of the following reasons except _____.
    a) ultrasound is a versatile technique
    b) ultrasound has proven widespread clinical utility
    c) ultrasound is considered highly toxic
    d) applications of ultrasound are growing considerably

510. **A, B, C, and D.** Tissue equivalent phantoms have characteristics that allow for the optimal evaluation of ultrasound systems. Thus, they are quite similar to soft tissue and are the preferred device to evaluate system performance.

511. **A and B.** The calorimeter and the thermocouple are two devices that use the principle of converting acoustic energy into thermal energy to estimate the energy in a sound beam. A hydrophone converts acoustic energy into pressure. A Schlieren uses the interaction of sound and light to evaluate sound beam profile.

512. **False.** The biological effects of ultrasound are studied extensively by scientists throughout the world. Under specific circumstances, bioeffects have been confirmed and are the subject of many discussions, investigations, and publications.

513. **False.** Ultrasonic bioeffects are substantial under certain conditions, including, but not limited to, therapeutic ultrasound and lithotripsy.

514. **True.** At the levels typical of current diagnostic imaging equipment, ultrasound is generally considered safe and nontoxic. The likelihood of significant and harmful bioeffects is extremely low.

515. **B.** Fractionation means "to separate a mixture into its ingredients," and it is not related to bioeffects. The other choices are potential mechanisms for acoustic bioeffects.

516. **C.** The AIUM examines bioeffects research because diagnostic ultrasound is a clinically relevant, popular, and a widely used imaging technique. Diagnostic ultrasound is not considered highly toxic or likely to expose patients to substantial risks.

517. Which of the following is not a valid rationale for establishing quantitative safety guidelines for the use of ultrasound?
    a) they would be applicable and appropriate for a majority of clinical situations
    b) the FDA would have a scientific basis for regulation of industry
    c) manufacturers would have guidelines for system design
    d) it would reduce concern of the patient community

518. Which of the following does not contribute to the difficulty in establishing standardized guidelines for the use of diagnostic ultrasound?
    a) the risks are small and hard to measure
    b) scientists are not convinced of the potential for bioeffects in man at standard diagnostic imaging intensities
    c) there are many varied ultrasound systems, and the exams performed are diverse
    d) it is difficult to measure the intensity of ultrasound beams in vivo

519. Some studies of ultrasonic bioeffects are performed in vivo. What does this term mean?
    a) observable in a living body   b) observations based on an experiment
    c) discernible in a test tube   d) perceptible in a plant

520. True or False? The mechanistic approach to the study of bioeffects and safety includes the identification of a theoretical construct that could produce an effect.

521. True or False? The empirical approach to the study of bioeffects and ultrasonic safety surveys data in hopes of finding a relationship between exposure and toxic effects.

522. True or False? A bioeffect identified via the mechanistic approach rather than by the empirical approach is most likely to have clinical significance.

523. True or False? The AIUM considers an ultrasound-induced biologic tissue temperature rise of less than 1^0 centigrade above normal body temperature as safe for clinical studies.

517. **A.** A major hindrance to the development of quantitative guidelines for diagnostic ultrasound exposure is ultrasound's use in many diverse clinical circumstances. For example, it is difficult to establish quantitative rules that apply to both endovaginal scanning for fertility studies and the evaluation of carotid artery stenosis.

518. **B.** The difficulty in establishing standardized guidelines include choices A, C, and D. Choice B does not contribute to the difficulty in establishing guidelines. First, it is not true that scientists are not convinced of the potential for bioeffects in man. Second, even if true, it would not hamper the development of such guidelines.

519. **A.** In vivo observations are made in a living body, for example, observations of increased tissue temperatures in patients following exposure to ultrasound.

520. **True.** The mechanistic approach is based on developing a theory as to how something happens. If the theoretical mechanism is correct, then great insight into the system and its component factors is achieved. The development of a computer model based on the concept that ultrasound increases tissue temperature is an example of a mechanistic approach.

521. **True.** An empirical study is one in which investigators look for a relationship between two events without necessarily understanding the fundamental cause and effect that relates the two factors. One example of an empirical study is that of examining the possibility that changes in fetal birth weight and ultrasound exposure are associated. Note that no specific mechanism for this relationship is postulated.

522. **False.** A bioeffect identified only by a mechanistic methodology is not likely to have greater clinical significance than a bioeffect identified empirically. The most clinically relevant bioeffect would be identified both mechanistically and empirically.

523. **True.** AIUM guidelines state that an exam is considered free of the potential for thermally induced bioeffects if the tissue temperature is within 1° C of normal.

524. According to the AIUM, at what in situ tissue temperature is there danger to a fetus?
    a) 100° C
    b) 98.6° F
    c) 1° F above normal body temperature
    d) 41° C

525. Of the following choices, which variable is considered the most important for the sonographer with regard to bioeffects?
    a) pulse repetition frequency
    b) frequency
    c) duration of the study
    d) imaging mode

526. Which of the following ultrasound beams has a characteristic most likely to cause temperature elevation in soft tissue?
    a) strongly focused
    b) medium focused
    c) unfocused

527. Which of the following statements regarding cavitation is true?
    a) it has never been observed in any biologic media
    b) stable cavitation relates to oscillating bubbles whereas transient cavitation relates to bursting bubbles
    c) it is a nonlethal bioeffect produced in animal experiments
    d) waves with peak pressures of less than 100 MPa can never induce cavitation

528. Research has indicated that cavitation _____.
    a) never occurs
    b) cannot occur with short pulses
    c) can be lethal to living things
    d) effects are purely theoretical

529. What is studied in epidemiology?
    a) large groups
    b) the prevalence of disease
    c) acoustic bioeffects on the fetus
    d) in vitro effects

530. Many epidemiologic studies of in utero exposure to ultrasound have concentrated on all of the following findings except _____.
    a) head circumference
    b) birth weight
    c) cancer
    d) structural abnormalities

524. **D.** Elevation of fetal tissue in excess of 41° C is considered potentially harmful.

525. **C.** The exam duration is most significant because it is directly controlled by the sonographer. Exams should always be of high quality and should provide relevant diagnostic information, with a secondary goal of minimizing output power level and time exposure to the patient.

526. **C.** An unfocused beam is most likely to cause temperature elevation in soft tissues. Temperature elevation with unfocused beams results from the broad area over which the beam spreads. With narrower, focused beams the area insonated by the beam is smaller and therefore is less effective in heating a significant mass of tissue.

527. **B.** This is the principal difference between the two forms of cavitation: stable cavitation means that the gas bubbles in the tissues are rhythmically swelling, whereas transient cavitation means the gas bubbles burst.

528. **C.** Cavitation resulting from ultrasound exposure is lethal to the fruit fly. Cavitation has the potential for significant and harmful bioeffects under specific circumstances.

529. **B.** Epidemiology is a division of medicine that is devoted to the study of the prevalence of diseases or pathology in a defined population.

530. **A.** Many epidemiologic studies on the effects of ultrasound have been reported. The head circumference in infants is the least studied parameter following in utero exposure to ultrasound.

531. What does the statistical power of an epidemiologic study indicate?
    a) how conclusive the results are
    b) agreement on the study's conclusions by the medical community
    c) the probability that the results are valid
    d) the number of observations required to yield a statistically valid result

532. What is the highest SPTA intensity of an *unfocused* ultrasound wave when there have been no observed bioeffects?
    a) 1 mW/cm^2 b) 100 mW/cm^2
    c) 1000 mW/cm^2 d) 1 W/cm^2

533. The upper limit for the SPTA intensity of a *focused* ultrasound wave when there have been no observed bioeffects is _____ .
    a) 1 mW/cm^2 b) 10 mW/cm^2
    c) 100 mW/cm^2 d) 1 W/cm^2

534. True or False? One reason why focused ultrasound beams with low intensities are less likely to cause bioeffects is that a focused beam is less efficient in heating a large mass of tissue to a critical temperature.

535. True or False? Focused ultrasound beams are considered less likely to create bioeffects because the beams will strike fewer gas bubbles that could potentially cavitate.

536. Certain studies of bioeffects are performed in vitro. What is the meaning of the term in vitro?
    a) visible in a living organism   b) observations based on experiments
    c) discernible in a test tube   d) perceptible in a living human being

537. Which component of an ultrasound system is most likely to expose a patient to risk?
    a) the CRT   b) the electric cord
    c) the pulser   d) the transducer

538. True or False? The AIUM suggests that in vitro research confirming bioeffects is valuable and valid. Their results are significant and should be directly applied to the clinical arena.

531. **D.** The statistical power informs researchers of the number of patients that must enroll in the study in order to yield a statistically significant conclusion. From this, one gains insight into the length and the resources required to complete the study.

532. **B.** There have been no independently confirmed significant bioeffects in mammalian tissue exposed to unfocused ultrasound with intensities below 100 mW/cm^2.

533. **D.** There have been no independently confirmed significant bioeffects in mammalian tissue exposed to focused ultrasound with intensities below 1 W/cm^2.

534. **True.** With focused beams, only a small mass of tissue is heated, and the heat flows rapidly to cooler neighboring tissues. There is limited accumulation of heat energy, and the temperatures tend to stay below critical levels.

535. **True.** Small gas bubbles situated in tissues serve as sites of cavitation. When a beam is focused, its diameter is smaller and is less likely to strike these cavitation nuclei.

536. **C.** An in vitro experiment is performed out of the body in an artificial environment. The observations may be described as "in a test tube."

537. **D.** Although the risk is small, the transducer exposes the patient to the greatest risk during an ultrasound scan. However, the risk will be substantially greater if the system is not functioning properly or is being used irresponsibly.

538. **False.** Although the reports of in vitro experiments are extremely important, it is difficult to assess their direct clinical significance. Further studies should be performed before applying these conclusions to the clinical setting.

INDEX